Flying off
RATTLESNAKE
MOUNTAIN

The Amazing Story of Pioneer Inventor-Aviator,
Micajah Clark Dyer

SYLVIA DYER TURNAGE

TURNAGE PUBLISHING CO., INC.
805 Low Gap Rd.
Blairsville, GA 30512
sturnage@windstream.net

Cover design by Melinda De Ross: http://melindadeross.com/

Library of Congress Cataloging-in-Publication Data
Turnage, Sylvia Dyer.
   Flying off Rattlesnake Mountain, the amazing story of pioneer inventor-aviator Micajah Clark Dyer – second edition.
1. Turnage, Sylvia Dyer—Flying off Rattlesnake Mountain, the amazing story of pioneer inventor-aviator Micajah Clark Dyer, first edition 2017. 2. Turnage, Sylvia Dyer--Choestoe Songs, 1993.
3. The Legend of Clark Dyer's Flying Machine, 1994.
4. The Choestoe Story, first edition 1995, second edition 2017.
5. Financially Savvy, first edition 2005, second edition 2015.
6. Georgia's Pioneer Aviator Micajah Clark Dyer, 2009.
7. Mountain Gold, 2015.

Library of Congress Control Number: 2017917724

International Standard Book Number: 978-1-880726-39-6 (hardcover)
International Standard Book Number: 978-1-880726-40-2 (softcover)

This book is dedicated to all of the descendants
of that remarkable man, Micajah Clark Dyer,
a genius who invented and flew an aircraft
in the 1800s, giving him the distinction of being
first to accomplish controlled flight.

Acknowledgments

I am very grateful to my husband Billy and our children for their unflagging support, encouragement and help in putting together this book. Without them, I am quite sure I would never have finished it.

A special thanks to my husband for most of the pictures that are included in the Appendix, to my daughter Karen and my dear friend Dottie Honea for editing/proofreading the chapters as I wrote them, and to my son Andrew for his assistance in getting the manuscript arranged for printing. Also, special appreciation to Ethelene Dyer Jones, Kenneth Akins, Watson Dyer and other historians whose research provided the foundation for an understanding of life during the 1800s as it was experienced in the North Georgia Mountains

Disclaimer

Names used in this narrative nonfiction novel are generally fictitious, except for Clark Dyer's immediate family, some public figures living during that era, and a few based on long-time legends connected to Clark and his family. Similarities of all other persons mentioned in the novel to persons living or dead is purely coincidental.

I wisdom dwell with prudence,
and find out knowledge of witty inventions.
Proverbs 8:12

CONTENTS

Foreword

Part 1
1833 - 1838

Part 2
1839 - 1863

Part 3
1864 – 1891

FOREWORD

MAYBE WHEN YOU THINK about the life of Georgians during pre-Civil War days you envision large plantations with long rows of cotton and lazy days of young folks partying. But people in the Southern Appalachian Mountains knew a far different way of life during, throughout and following the Civil War. They were hardy folks who earned their living by the sweat of their brow. They valued higher education, but schools were scarce throughout the region, so many strived to learn through every alternative means available to them.

The land into which the Dyer family came in 1833 had been surveyed and platted by the government only a year earlier to prepare for conducting the last land lottery. It was by this means that men and women who were successful drawers in the lottery gained title to a parcel of the land that had for centuries before been occupied by Cherokee and Creek Indians. The land was mostly steep, rocky and wooded. It had to be tamed with primitive tools, and the new owners had to be self-sufficient and industrious because it was the only way they could survive. Neighbors were usually miles away, dealing with their own hardships, so if the family couldn't do it themselves, they had to do without.

There were no cameras or newspapers available in this backwoods country in those days. Although the mountaineers were able to get newspapers and books about events outside their home area when they went into the larger towns to trade for supplies, outsiders didn't hear much about anything that happened inside this rough and secluded country. And the residents were far too busy with the work necessary for feeding and clothing their families to write about anything that was going on around them. Indeed, it likely did not occur to them that anyone would have an interest in what they were doing.

Sometimes, into a setting such as this, a genius is born, a person who takes the limited education he has and extends it beyond the boundaries of ordinary imagination and creates things that men and women of lesser intelligence cannot understand or appreciate. Because of a lack of understanding, they belittle the genius for spending long hours on what seems to them to be only frivolous activity. And that is how it was with Micajah Clark Dyer.

In essence, this is a true story. It depicts life as it was lived by the North Georgia pioneers and how they coped with hardships of war and economic upheavals. But its primary purpose is to give an account of how Clark came to have a remarkable theory, the roadblocks he encountered as he toiled to bring his vision to fulfillment, and the ridicule he sometimes suffered from family and neighbors because they could not understand him and his passion for flying. It is based on information gleaned from every available source—old newspapers, census reports, family recollections, and historical records, some of which are included in the Appendix of this book.

My hope is that you will come away with a new appreciation for the strong men and women who settled this part of the country, and especially that you will see just how remarkable were the accomplishments of Micajah Clark Dyer.

SYLVIA DYER TURNAGE

Part 1

1833 - 1838

Happy is the man that findeth wisdom,
and the man that getteth understanding.
For the merchandise of it is better
than the merchandise of silver,
and the gain thereof than fine gold.
Proverbs 3:13-14

1. Coming to Choestoe

"Whoa," Elisha Dyer called to his horse as the wagon topped the ridge at Tesnatee Gap. Even in the cool, early spring breeze, the muscular horse had sweat trickling down his neck and he was breathing hard from pulling the loaded wagon up the south side of the Appalachian Mountain range. It had been two hours since the little caravan's last stop, and the winding trail they traveled was barely wide enough for the creaky covered wagons to pass between the trees.

Elisha and eleven-year-old Clark walked to the front of the wagon and looked toward the panorama lying before them. The boy closely watched his grandfather's weathered face as he scanned the vast gorge and the rumpled mountains stretching to the north. Elisha's blue eyes showed excitement and determination as his gaze swept the rugged terrain between him and the horizon. He had moved his family many times in the past twenty years, and this time he had chosen the location for their new home carefully, knowing it would be the last move for him.

Clark had heard his grandfather's stories about Choestoe for months now, for he had made many trips across the mountain to prepare the property for bringing the family over. He told them about the Indians who lived nearby and the itinerant gold miners with tall tales of their experiences. Clark was eagerly anticipating living in a place such as this.

Five other loaded wagons crested the mountain range and came to a halt behind Elisha. The drivers walked to the front to survey the path ahead.

"Holy smoke!" cried Bluford, looking down the steep mountainside. "We'd better make sure our brakes are working before we start down that trail!"

"Yep, son. I nearly let my wagon get away a few times before when I traveled through here. You better make sure you brace yourself and keep a good hold on the brakeboard. The next mile is going to be rough for all of us."

Elisha leapt nimbly to the ground and began checking the brake boards on each of the wagons. A tall, slender man, his agile steps belied his age of almost 50 years.

A baby's cry drifted from Elisha's wagon. He quickly walked back and looked inside where his wife Elizabeth was holding their nine-month-old son Lump, flanked on each side by their little daughters, Melinda, age six, and Matilda, age three. They were sitting on a featherbed spread over some of the household furnishings that they were hauling to their new home in Choestoe.

"Pa, will there be apples on the trees in Choestoe?" Melinda asked, squinting her eyes against the setting sun as she raised her little bonnet-framed face upward toward him. She loved the apple tree in the back yard at the home they were departing, as well as the swing hanging from one of its big limbs, and she wasn't happy to be leaving it behind. She had spent many contented hours playing beneath that tree.

"We won't have an apple tree at our new home right now, but I did see an orchard on the road going down to the trading post. Don't worry. I'll get you some apples, and I'll plant two big apple trees in your new backyard," Elisha said comfortingly. He knew Melinda would miss her older sisters who were now married and had babies of their own.

They had all lived near each other in Habersham County under the shadow of Yonah Mountain, and hardly a day passed that they didn't visit back and forth between their homes. But two of his married daughters, Amy and Mary, weren't moving with them to Choestoe. Now they would seldom see each other because the trip over the mountains would take too long and it held many dangers for a lone traveler or a small group. Renegade Indians, bitter over the white man's intrusion into their hunting territory, often attacked travelers along the path, stealing their horses and whatever else presented a novelty to them. Panthers, bears and rattlesnakes had their homes in the mountains, and they too were a threat.

"Pa," Cager called. "Clark and Jimmy are playing instead of helping get the wagons ready for the trip down the mountainside." Clark had made a peashooter from a hollow willow stalk, and finding a fox grape vine heavy with green grapes, he had filled his pockets. Now he was taking turns with Jimmy shooting at a toad that was trying to leap out of harm's way.

"Let them work off a little energy, Cager," Elisha said. "They have been good travelers all day." He stepped over and took the peashooter from Clark.

"Son, that's pretty good workmanship," he said as he looked it over. Somehow, no matter where Clark happened to be, he always found something he could fashion into a toy for himself or one of the other children, and Elisha had taken note that his grandson's craftsmanship was quite expert for a boy his age.

Cager and Clark had been given the same name at their births, Micajah Clark Dyer, but that seldom caused any confusion because everyone called the older boy Cager and the younger one Clark. Cager was Clark's uncle but he was only five years older. Clark had been born to Cager's sister, Sallie, when she was only 18 years old. Elisha had been adamantly opposed to Sallie marrying the man, John Meyers, who had fathered her child. Their neighbor, Eli Townsend, was smitten with Sallie, and Elisha was certain he would make a much more suitable husband for her.

Sallie had married Eli when Clark was just a baby, but Elisha and Elizabeth had already become very much attached to the child and they convinced Sallie that it was best for them to keep Clark and raise him with their youngest child, James, whom they called Jimmy and who was nearly two years old at that time. The boys were as close as identical twins; wherever you saw one you could be sure the other was nearby.

"I've decided to camp here for the night. There won't be a place level enough to do it as we go down the other side," Elisha said to Bluford. "Then, we can start early in the morning and should be able to reach Choestoe well before nightfall."

"We're going to camp here," Jimmy called as he ran past the other wagons, and the women in their long dresses began gingerly climbing down to avoid snagging their skirts on the rough wagons. As they were unloading pans and supplies for making the evening meal, they called to the boys to gather wood from the surrounding area to make a fire.

"But don't wander far and get lost," Elizabeth warned them. "Bluford, you keep a watch on them boys. Lord knows, if they get to exploring in the woods they'll completely forget what they're supposed to be doing."

The men began tying the horses, oxen and other farm animals to trees away from the little clearing where they would be spending the night. They pulled bundles of hay from one of the wagons and spread it for the animals to eat. Since Elisha had traveled through the area before, he knew the location of a nearby spring. He took a five-gallon bucket and a large gourd dipper from the wagon and started out to get water for cooking and drinking.

"Can I go with you, Pa?" Matilda asked as she took hold of his free hand.

"Sure you can, Sis. I might even let you carry the bucket of water back for me," Elisha teased.

"I will carry it for you, Pa," Matilda said, totally unaware in her three-year-old mind that she was promising something she couldn't deliver. Elisha smiled at the little tow-headed girl squinting her bright blue eyes as she looked up at him. He felt humbly grateful that his large family of six boys and six girls had been blessed with good, strong minds and bodies.

As Elisha was returning with the bucket of water, Matilda chattering away at his side, he heard frantic calls from the campsite, "Jimmy! Clark! Where are you?"

Hurrying to the others, he asked, "Which way did they go?" Elizabeth, wringing her apron around her hands in anxiety, pointed up toward a small trail along the mountainside.

Elisha set the bucket down, patted Elizabeth on the back, and quickly started up the mountain. Nightfall was fast approaching and he knew the boys would be nearly impossible to find in the darkness. He whistled through his teeth as he went through the woods, knowing the sound would carry better than yelling. After a series of three short shrill whistles, he listened intently to see whether he could hear any answer from the boys.

Finally, when he had gone quite some distance along the trail, he heard an answering whistle down the mountainside. Turning off the trail, he headed toward the sound, continuing to give the three short whistles and continuing to get three in return.

Coming upon the boys, he said sternly, "Now, you boys know better than to go off and get yourselves lost. What if a panther had found you before I did?"

"We didn't mean to, Pa. We just kept seeing good sticks for firewood and kept going down the mountain till we couldn't remember where the trail was anymore," Clark said.

"What have I told you about always noticing landmarks and directions when you're in a strange place?"

"I know, Pa. I will remember next time to look at the trees and see which side the moss grows on and break branches on the bushes to follow back." He always felt considerable remorse when he disappointed his grandfather. No one else listened to him so closely or gave him such good advice and encouragement.

"Let's get the wood and head back to camp. We must let your Ma know you're okay. You nearly scared her to death, disappearing like that. Shame on you!"

When they got back to the camp, the women had warmed the food they brought over the little campfire. They had prepared well for the trip before leaving Habersham County and were serving a good meal to the family in tin plates. Clark thought he had never tasted beans, sweet potatoes and cornbread that were better than this.

The rest of the evening went without incident and the little group settled down for the night around the campfire, except for the babies and toddlers who were bedded down in the wagons.

Next morning, they arose early and prepared the wagons for their descent from Tesnatee Gap into the valley that would be their new home. Everyone who was old enough to walk would do so. It would be safer than riding in the wagons, which might lose their brakes on the way down and careen into the gorge. Elizabeth, with baby Lumpkin, was riding one of the mules.

They stopped only once for a mid-day meal during their descent, and just as the sun was beginning to set, the caravan arrived at a beautiful valley with a bold creek running through it.

"This is it," Elisha announced proudly. "What do you think of the place?"

Murmurs of approval came from the family members as they gazed across the pristine land. They had received the property through land lottery drawings. Their original drawings had been for land in Kentucky, but they had sold those properties and purchased this land, which they heard about from travelers who passed through their property in Habersham County.

"It is called Choestoe," said Elisha. "According to the Indians, it means 'land of the dancing rabbits.' Actually, I've heard many white men say that they've seen the rabbits dancing here. I think it's going to be a very good place for us to live."

The men circled their wagons and began setting up camp to spend their first night in their new homeland. It felt like a place where life was going to be good to them.

2. Life in the New Land

Next fall found the Dyer family with a new home, finished and occupied. The other families who came with them had also built their homes, and all were settling into a comfortable routine. Crops had flourished in the new ground they had tilled and planted, and the root pits were stocked with sweet and Irish potatoes to take them through the coming winter months. Bags of apples, dried by the womenfolk, were hanging in the smokehouse. Onions, leather britches and peppers were hanging on yarn strings from the nails driven into the wall in the back hallway of the house. Elizabeth had pickled corn, beans and kraut in large churns, knowing that her sizeable family would need a lot of food before crops could be harvested again.

The men were watching some wild hogs that were feeding on acorns and chestnuts in the mountains above the settlement. Soon they would entice them into pens, where they would be fattened on corn and table scraps for a few months before being slaughtered and cured after the weather turned cold. The chickens they brought to Choestoe had multiplied, despite a few losses to foxes and hawks.

The afternoon sun was showing up the brilliant red maples and golden beeches. A gentle breeze swept across the mountainside, fluttering the leaves and whispering through the pines. Elisha sat on the front porch of the log house they had built, listening to Cane Creek splash over the rocks as it wound its way through the valley toward Nottley River. He was sharpening his scythe with a whetstone, and he rocked back and forth as he rhythmically ran the stone across the blade. As he slid the stone repeatedly across the blade with a swishing sound at each stroke, he looked completely in harmony with the nature around him.

"Pa, Pa!" shouted Clark and Jimmy as they suddenly rounded the corner of the house at breakneck speed. "Come quick! A fox has jumped the fence and is killing all the chickens!"

Elisha leapt to his feet, grabbing his bow and arrow pouch from the knob on the porch post. As he reached the chicken pen, he deftly fitted an arrow into the notch of the bow and took aim. The arrow found its mark in the fox's neck and it rolled over, kicking up the dust as it struggled to regain its footing. The predator's efforts were futile and soon it lay still.

Elisha and the boys jumped over the fence to see how many chickens they had lost. They counted four, when one of the hens stood up unsteadily, shook her feathers and ran staggering into the chicken house.

"Well," said Elisha, "We'll have chicken and dumplings for supper tonight. It's just too bad that we can't eat that rascally fox."

They picked up the three dead chickens and took them to the chopping block. Elisha picked up his axe and severed their heads, tied their legs together, and hung them on a tree limb so their blood could drain.

"Go tell Ma to get some water boiling 'cause we're bringing her some chickens to scald and pluck."

Clark and Jimmy ran toward the house, glad for the excitement of something different going on. Elizabeth was sitting at the spinning wheel with yarn rapidly filling the spindle she held. Melinda and Matilda sat beside her carding the wool that Elisha had sheared from his sheep back in the spring. Lump was crawling in and out of a big box Elizabeth had given him to play with while she worked.

"An ol' fox just killed three of your chickens!" Clark said as he came into the house. "Pa said for you to get water boiling so

you can get them ready to cook for supper tonight. He wants you
to make chicken and dumplings."

"Oh, my goodness!" Elizabeth said. "Pa is probably glad the
ol' fox killed them. I never saw a man who loved chicken and
dumplings better'n him. You boys run to the creek and get two
buckets of water."

She laid her spindle aside and went quickly into the kitchen.
She added some kindling wood to the embers in the fireplace and
blew on them. As soon as they flamed, she added larger wood,
and by the time the boys returned with the water, she had a
blazing fire going and was ready to fill the big iron pot hanging
on the hook mounted in the center of the fireplace.

"Clark, get me the butcher knife from the kitchen shelf,"
Elizabeth called before the boys could get out the door and head
back to their wood-splitting chore.

"Ma, how are you gonna reach things on that shelf when all
of us are gone from home?" Clark smiled teasingly at her.

"Don't worry 'bout me. I can drag a chair over there and climb
up anytime I have to." She was fiercely defensive that her
diminutive size didn't hinder her from doing everything anyone
else could.

As the boys stepped back outside, they saw a girl coming up
the path toward the house. At first, they thought she was an Indian
girl because she wasn't wearing a bonnet and her arms were bare
and brown. But as she came nearer, they saw that it was Morena
Owenby whose family lived two miles away in the Indian village
beside Wolf Creek. The Owenbys were one of several white
families who had lived among the Indians for a half-dozen years
or so. They had adopted a lot of the Indian manners and lifestyle,
which caused them to seem strange to the newcomers. Some of
the families had inter-married with the Indians and, for the most
part, the new settlers, who were Scotch-Irish, felt they were
superior to them.

Morena came quickly up the path to the steps and stopped.
She looked shyly at the boys on the porch. "Is Miz Dyer here?"
she asked softly.

"She's in the kitchen plucking chickens," Clark said. "I'll tell
her you're here." He gave her a curious look before going inside.

"Ma," he said, coming into the kitchen, "Morena, the half-Indian girl from Wolf Creek, is here and wants to see you."

"Clark! That girl is not half-Indian! The folks around here say she is because she doesn't keep her skin covered when she's in the sun. She's just tanned and wears her hair and clothes like the Indians. Ask her to come on in here. I can't stop what I'm doing right now to go out there."

Clark returned and delivered the message to Morena, who gave him a slight smile as she came up the steps and went in the house.

Turning her head as she heard Morena come into the room, Elizabeth said, "Howdy, Morena. Is there something I can do for you?"

"My Ma wants to know if she can come over here sometime and let you teach her how to spin and weave. She got a new spinning wheel and loom, but can't figure out how to use them." She added shyly, "I want to learn, too."

"Why, sure, of course. I'd be happy to teach you both. Tell her to come the middle of next week in the early afternoon when I am usually doing my spinning and weaving, and she can see how it's done. It's not hard work, but it does take some coordination and practice to get it right. She'll probably catch on to it in no time."

"We appreciate you for doing that, Miz Dyer. My Ma will be glad to do anything she can for you, too. We'll be here next week for you to start teaching us."

Morena left quickly and went down the path leading to the wagon trail without even glancing toward Clark and Jimmy at the woodpile. But they were inquisitively watching her. Unlike the teenage girls in their settlement who pinned their hair up into buns, braids or twists, Morena's dark hair was loose and fell to the middle of her back. Her Indian-style dress only came to the middle of her calves, and she was wearing moccasins.

Shaking his head, Clark said wonderingly to Jimmy, "Ma said she ain't half-Indian, but she sure looks like it to me."

"I'll say she does," Jimmy agreed. "Do you think her family will move away with the Indians when the government makes them go?"

"I doubt it. The Indians around here don't make any difference between their race and ours, but others away from here won't accept them in their villages. I guess the whites who have been living here with them are going to miss them a lot when they go," Clark said quietly.

Actually, he had thought a lot about how awful it would be to have the government round up a race of people who had lived in a place all their lives and send them to a faraway land. He knew some of the whites didn't consider Indians as having the same feelings that whites did, but he was sure there wasn't any difference. He admired the way the Indians were fiercely loyal to their tribes, and he noticed they didn't brag a lot about themselves or make fun of each other as his race often did.

As the sun dropped low, the boys left off splitting wood and went to round up the cattle and horses that had not already come to the barn for their evening feeding. Lige and Cager rode up, dismounted their horses and threw the reins to their younger brothers. They had worked all day helping Zeke Kinsey build a barn. Zeke didn't have any sons, and he had called on them to lend a hand in putting up the rafters.

Turning to go to the house, Cager suddenly let out a bellow and clutched his arm. The startled brothers all turned to see what was wrong and spotted a rattlesnake coiled on the worktable against the wall. It had bitten Cager's arm as he passed by. Lige grabbed a hoe leaning against the side of the barn and dragged the snake onto the ground with it. The snake was striking wildly at him, but he kept his distance. Seeing it had no chance of escape, the snake tried to bolt to the bushes. Lige raised the hoe and brought it down with a strong, swift blow that severed the snake's head.

"Hurry to the house and get Pa to take care of the bite, Cager," said Lige. Turning to Clark and Jimmy, he said, "You know rattlesnakes travel in pairs, so another one is somewhere around here. You boys better keep your eyes open or you'll be the next ones screaming."

Darkness was coming fast and the boys hurried to get the animals fed, fearing that any minute they might feel poisonous fangs sink into an arm or leg as they reached into the haystack and corncrib to get feed. But with great relief they finished their work without incident and rushed to the house to see how Cager was doing.

Elisha had made a crisscross cut over the spot on Cager's lower arm where the snake had bitten it, and he had sucked out as much of the contaminated blood as he could. He poured whiskey over the wound and bound the arm with gauze. Then he tied it to a splint. It was already swelling and Elisha knew he would have to keep a close eye on the binding to keep it from becoming too tight.

Elizabeth took Cager's arm gently and led him to bed where she had placed folded quilts to keep his upper body elevated above the wounded arm. As he settled into position, she laid a cool, wet towel on his forehead. "You'll be okay, son," she said, stroking his shoulder. "Try to keep from fretting and breathe deep. You don't want your heart to be racing with all that poison in you. I'll bring you some chicken and dumplings in here directly. You ought to try to eat a little something."

Elizabeth went back to the kitchen and began putting food on the table for the very subdued family. They knew that a snakebite was serious and that complications could bring death to the victim. They would have normally been diving into the special treat of chicken and dumplings, but now they ate slowly and no one was talking. Only Lump, too young to understand the urgent situation, played on the floor, laughing as he pushed around the little wooden cart that Clark had made for him.

3. The Long Trip

Clark woke early as the big Rhode Island roosters began crowing outside the window. He shivered as he slipped out of the warm bed he shared with Lige and Jimmy. Pulling on his cold britches and grabbing socks and brogans, he headed to the kitchen to start a fire. Ma would be up shortly to fry ham and eggs and make a hoecake for him and Pa before they left for Gainesville. They were taking a wagonload of grain, honey and hides to the merchants in town. They would barter these for the goods they were unable to raise on their farm—coffee, salt, baking soda—and cloth, shoes, kerosene oil and turpentine.

Clark had never been to Gainesville, and he was very excited that Pa had chosen him to go along on this trip. The trip would take four days of travel. They would sleep beside their wagon the first night as they crossed the mountain, and stay the second night with part of the family they had left behind in Habersham County. They planned to stay the last night in one of the stagecoach inns that were dotted along the wagon trail. Clark had heard stories about the inns, and he hoped he would get to meet some of the colorful characters such as those he had heard his uncles and neighbors tell about. His Uncle Will never missed a chance to tell about the gold miner he shared a bed with in Auraria one cold winter night.

"Lord! That man's snore would wake the dead!" he'd declare. "But the worst of it was that one of the men sleeping in the bed across the room from us would get mad and throw a boot at him. Well, when one of the boots would hit him, he'd sit up in bed and peer around in the darkness. Still more asleep than awake, he'd whimper, 'Minnie, don't hit me agin and I swear I won't never tetch another drap o' likker.'" Uncle Will would slap his leg, throw his head back and roar with laughter at the memory.

Clark was also looking forward to meeting their relatives who lived in Gainesville. He had heard about their fine house and the big library with dozens of books on every subject under the sun. He and Pa would stay two nights with them while taking care of business.

Elizabeth came into the kitchen pulling on a jacket against the morning chill.

"Clark, don't let your curiosity get you lost while you're down there in Gainesville. Your Pa will have his mind fixed on his business, and he might not be paying any attention to where you are."

"I'm not a kid, Ma!" Clark said indignantly. "I won't get lost."

"There's a lot to see in a town like that, and I'm just warning you that you need to keep within sight of your Pa."

Elisha came into the kitchen and heard Elizabeth's cautioning words.

"Now, Lizzie, don't go worrying about Clark. He's getting old enough to pay attention to what he's doing. I'll have him with me most of the time anyhow. Don't be fretting yourself about us while we're gone. You'll have your hands full looking after the little ones here."

Elizabeth handed him a piece of paper. "See if you can find some pretty calico while you're in Gainesville. I'm nearly out of cloth for making dresses for the girls. And the boys need new shirts. See if you can get some broadcloth, too. Here, I wrote down what I need if you're able to come across anything."

Clark picked up the milk bucket and headed to the barn. He knew he would have to hurry to get both cows milked by the time

Pa hitched up the horses and loaded their trunks on the wagon that had already been loaded with the goods to be sold. Pa worked quickly and would eat his breakfast hastily, then want to start the trip immediately.

As he came into the kitchen with two large buckets of milk, he saw that he had guessed right; Pa was just getting ready to sit down. Clark quickly covered the buckets with a cloth. Elizabeth would strain the milk later into gallon jars that Lige or Jimmy would take to the springhouse where it would keep cool until they were ready to use it.

Clark washed his hands and sat down at the table. He and Pa devoured the meal Elizabeth set before them, and rose to start their trip. She handed them buckets filled with food to eat along the way.

As they climbed into the wagon, sunlight was just beginning to spread orange and golden rays across the sky above the Blue Ridge. Gossamer fog was rising from the valleys, and the cool, moist air felt pleasant to Clark as it blew across his face and rustled his hair.

"Giddap," Pa said to the horses as he flapped the reins. The big chestnut-coated Belgians bowed their necks and bent their legs, and the heavy wagon began rolling down the trail. Two red hounds started to follow behind them, but Pa snapped his whip above their heads and said sternly, "Stay, Boss! Stay, Jack! You have guard duty here while we're gone." Actually, now that Bluford, Cager and Lige were nearly grown, he wasn't too worried about leaving for a few days. They knew how to run the farm and household while he was away.

By the time they rode past the last little Indian settlement at the headwaters of the Nottely River, before starting the long, steep ascent to Tesnatee Gap, the sun had crept over the Blue Ridge, and the Indians were already going about their daily work. The tribal men had adopted the ways of the English settlers and now, instead of spending their days hunting and fishing, they worked in the fields alongside the women. All of the villagers stopped what they were doing to watch the wagon pass. They answered Elisha's nod and wave with the same greeting sign.

It was nearly dark when they reached Tesnatee Gap, and they stopped to make camp for the night. After eating the meal Elizabeth had sent with them, they fed the horses and settled down in their blankets. The sound of crickets and cicadas echoed across the mountains, and a cool night breeze blew across the campsite. Soon Elisha and Clark were sound asleep.

Waking the next morning, they felt stiff from sleeping on the hard ground but were glad for no disturbance through the night. They broke camp, hitched the horses to the wagon and started the descent to Habersham County. The trail was not quite as steep on the south side of the mountain range, but still it required them to keep a strong hold on the brakeboard to avoid having the wagon get out of control.

As they reached Amy and John's farm that evening, the sun had already set and they were relieved to have the hardest part of their trip behind them. Driving the horses up to his daughter's house, Elisha called out, "Is anybody home?" Both Amy and John rushed out to greet them.

"My goodness! It's sure good to see you, Pa," Amy said, hugging her father warmly.

She put an arm out to Clark and said, "Look how tall you're getting! I bet you're ready for some supper."

"Yeah, we are," Elisha said. "But don't go to a lot of trouble for us. Just put out whatever you had left."

John shook Elisha's hand and patted Clark on the shoulder. "Amy sure has missed having you live close to us. She talks about you real often," he said. "Let me take the horses to the barn."

"How's Ma and the kids?" Amy asked as they went in the house.

"They're doing very well," Elisha said. "But a rattlesnake bit Cager on the arm about ten days ago. You should see what a big, black arm he's got right now. He's blessed he didn't die from it. They all said to tell you hello and give you a hug for them. Ma sent you some things in this bag that you can open before we go to bed."

"I guess Lump and the girls have grown so much we wouldn't know them."

"Prettiest little girls you've ever seen! Melinda is going to school and has really taken a liking to books. She plays school with Matilda so much that she can read a few words, and she's just now five. Lump's got big enough to get into everything. He keeps your Ma jumping. How are you and all the folks on this side of the mountain doing?"

"We've been doing fairly good. John and his daddy put in their crops together this year. His daddy is getting pretty feeble, and John only lets him do a little of the plowing in the cool of the morning. I take the kids and go over and help his mama with the washing every Monday. She's not able to do it herself anymore."

They went into the kitchen and Amy began dipping their food from the pots sitting on grates in the fireplace. She had cooked pinto beans with ham, sauerkraut, cornbread and blackberry cobbler. As she poured each of them a glass of buttermilk, Elisha thought how much she was like Elizabeth, small and moving quickly as she put their meal on the table.

As soon as they finished eating, Elisha brought in the bag for Amy. She first pulled out a fringed blue cape that Elizabeth had knitted for her. She spread it across her lap and stroked it. "Ma knows how I love blue. Hug her for me and tell her I will think of her every time I wear it."

Next she took out a brown toboggan Elizabeth had knitted for John. Then there were coats she had made for all of the children. Finally, in the bottom of the bag she found an olive-colored woolen blanket that Elizabeth had woven on her loom. It was heavy enough to keep them warm on the coldest winter nights. She knew her mother had boiled walnut hulls to make dye for coloring the woolen yarn she used to weave the blanket.

"Oh, Ma is so sweet to send us all these things!" Amy said. Suddenly, she bowed her head and began to cry at the thought of her mother far away on the other side of the mountain. The men remained silent and uncomfortable at her display of emotion.

"Amy, cheer up a little," said Elisha finally. "Clark, why don't you get the box with the things you made for the kids and let her see them."

Clark brought the box to her, and with a weak smile she opened it. Her smile grew wider as she found whistles he had made from willow canes, whirligigs that spun at the slightest movement, and wooden carts with little wooden wheels.

"The kids will love these, Clark! Apparently, you're still using your spare time to make things that move. You know, you have a special gift for figuring out how to make such clever things."

Clark smiled at her and wondered if she would be surprised to know that the happiest hours he spent were those when he was creating things from whatever materials he had around. It seemed to him that his mind was always racing, wanting to know what caused things to work or not work, or to see if they would work better if he made them differently.

"We'd better hit the hay, I guess," said Elisha. "We'll have to get started early in the morning. I hope we can get to Auraria by tomorrow evening."

Amy lit a candle and led them down a narrow hall to the little back bedroom where they would sleep. After the past night of sleeping on the ground, the bed felt like a soft cloud to them, and they soon drifted into a deep and peaceful sleep.

4. The Stagecoach Inn

Clark woke to the sound of Amy scurrying around in the kitchen as she prepared breakfast. He turned over and found Pa missing. He scrambled out of bed, dressed hurriedly and stepped through the door, rubbing his hair down with one hand and tucking his shirttail in with the other.

"Good morning," Amy said with a cheery smile as he came into the kitchen. "Did you sleep good last night?"

"I must have," Clark responded. "I didn't even hear Pa get up. Where is he?"

"He's out looking over the place and seeing what changes we've made since he moved away," she said laughingly. "Sit down and tell me how the rest of the family is doing while I finish cooking breakfast for us. The letters I get from Ma and them are always welcome news, but nobody gives enough information for me to get a good picture in my head of how everybody is situated there in Choestoe. Who lives close to you?"

"Well, they're all pretty close to us," Clark said. "Sallie and Eli live about a mile up the creek from us. Hezekiah and Sara live about three miles up the trail towards Enotah Mountain. Lucy and James live about four miles going toward Track Rock Gap. Did Ma tell you that James was elected sheriff of Union County?"

"No! What does Lucy think about that? Is she worried that he might get killed?"

"No, I don't think she worries much about that happening. James says most of the arrests he makes are when the gold miners at Coosa Creek get to fighting over stolen mules or bags of nuggets."

"John's brother worked for a while at the mine over at Dukes Creek, but then he decided farming suited him a whole lot better," Amy said, as she began putting food on the table.

Elisha and John came in and sat down at the table with Clark. Amy brought in a plate of biscuits she had just taken from the oven to complete the meal, then she sat down with them. They bowed their heads and John asked God's blessing on the food and their families, for safety for Elisha and Clark as they traveled, and for forgiveness for their sins.

They continued sharing news about the families on each side of the mountain while they ate. As soon as they were finished, Elisha was ready to continue the trip. He gave Amy a hug, shook hands with John, and climbed into the wagon with Clark. With a slap of the reins on the horses' necks, they were soon rolling southward along the little wagon trail.

After they had traveled several miles, they saw a gate across the road up ahead. Elisha turned to Clark. "Reach under your seat and hand me my money pouch," he said. "This is a toll road and we have to pay at the gate to get to travel on it."

"Why would anybody charge us to travel on the road?" Clark asked.

"It's the fair thing to do, son. It takes a lot of work to keep up a road like this. It's not just farmers like us that travel through here. The stagecoach hauls travelers back and forth every day and the mailman makes deliveries to all the post offices through here. When the roads are wet, and especially when they are thawing in the spring, the loaded wagons rut them out really bad. Sometimes trees fall across the road, and somebody has to cut them up and move them out of the way. It takes a lot of maintenance. I don't begrudge the man his toll charge. I'm glad to be able to travel without having to stop and clear the road."

Clark looked at Elisha appreciatively. It was a great pleasure to him for Elisha to explain how things such as this worked. There was a lot to learn, Clark thought, and he wanted to know about all of it.

Near noontime they came to a little stream with a large oak tree beside it. Elisha pulled the horses over and let down the reins for them to drink. Then he led them over to a grassy patch where they could graze and rest while he and Clark ate their lunch.

After eating, they stretched out on the grass and watched the crows circling overhead. "Pa," Clark asked, "Do you think it would be possible to build something that a person could fly around in?"

Elisha chuckled. "Now, what use would it be to a person if he could fly?"

"Well, you could go a lot faster and you wouldn't have to have roads to maintain for traveling. It must really feel good to soar in the air like that."

"Yeah, that is so," Elisha conceded. He stood up and started toward the horses. "We'd better get going, boy. We've still got a lot of miles to cover before night."

Now that the thought of flying had occurred to him, Clark couldn't keep the ideas from wildly bouncing around in his head about how this could possibly be done. Would a person need to make wings like a bird? What could it be powered with? How could it be controlled? He was certainly going to see if he could build a flying toy when he got back home.

Hungry and tired, they reached Auraria Inn just before nightfall. A stagecoach and several covered wagons were pulled under the trees, and Elisha guessed that they would be lucky if the inn still had a bed for them.

They went inside and were relieved to find that there was one bed left for hire. Elisha paid for it and went back outside to pull his wagon out of the way and take his horses around back to the corral and give them provisions. Returning, he and Clark went to the dining room and found a plentiful meal awaiting them.

A variety of folks were staying at the inn that night—several farmers, gold miners, a young husband and his wife, and an elderly woman accompanied by her middle-aged daughter. The men ate heartily, as did Clark, but the women ate sparingly, seemingly ill at ease among so many rough strangers at the table. As soon as they finished the small servings of food they had put on their plates, they excused themselves and went to their rooms.

With the women gone, the conversation began to get loud and coarse. One of the miners, who had obviously been imbibing already, pulled a jug of moonshine from his bag and offered it around the table. A grungy miner sitting next to Clark handed him the jug and said, "Ye look old enough to be able to handle some likker, boy. Hyere, have a drank."

Elisha reached across Clark and snatched the jug from the miner. Giving him a fierce look, he said, "I don't reckon it's any of your dang business deciding when a son of mine is ready to drink. If nobody else here wants a snort of this, I think the owner needs to put it back where it came from."

The innkeeper hearing the voices getting louder in the dining room came in and said, "Let me make something clear to you men. If you're planning to stay here tonight, you better not be drinking and causing a ruckus. I will put you out to sleep in the barn if you do."

Looking fiercely around the table at them, he asked, "Anybody got any notions of doing something other than minding your own business and letting the rest of the folks sleep peaceably around here tonight?"

Nobody answered him. He said, "Good," and stomped out.

After a brief period of quietness, the conversations started up again. One of the farmers, Thomas Self, was quite well-educated and mentioned that he taught seven grades in a little one-room school in his neighborhood. He was very proud that most of the parents took an interest in sending their children to school.

"They need to learn more than just a trade. They need a knowledge of literature, history, government and science," he said.

He expressed concern about the treatment the Indians were getting from the government. "The government officials have convinced the Indians to learn to farm instead of just relying on hunting for their food. They've taught them that they must live by the government's laws, telling them this is the way the two races of people can live together in peace. Every time the government has signed a treaty with them, they say it guarantees them that they'll keep all of their Cherokee land not ceded in the treaty 'as long as the grasses grow and the rivers flow.' But now they're getting ready to force them completely out of this territory despite all of the promises."

"Well," drawled one of the miners, "I reckin they oughter have sense enough to quit signin' treaties if hits a-causin' 'em harm."

"The treaties are being negotiated with just a handful of the Indian leaders, and the broad population doesn't know or understand what is taking place," Self said. "But, even so, it just isn't right to take away their land when they have relied on the promises that were made to them that they could remain here without further encroachment by us."

Elisha had heard this discussion a dozen times in the past, and he knew the men would not see eye to eye on the subject even if they talked all night. He stood up, nodded to the group and started towards the narrow stairs at the end of the room, with Clark close behind him.

Entering the room, they saw that the sloping ceiling was low and nails from the roof shingles were sticking through, exposing about a half-inch of the sharp ends. There were four beds that would sleep eight of the men tonight. A candle on a little table near the door was dimly lighting the room, and they made their way between the beds to the back.

"You sleep on the rear side, Clark. I'll have a little more ceiling space on the front side to keep my head from getting nicked by those wicked looking nails," Elisha said with a wry grin.

The bed had ropes woven across the frame and a straw-filled mattress lay on top. The pillow ticks were filled with cotton. The sheet, pillowcases and homespun blanket appeared to be clean, and Elisha hoped for a decent night of rest here.

Clark settled down on the backside of the bed and found it to be fairly comfortable. He wished he could drift off to sleep, but he had seen so many things that day his young mind was turning them over and over. He also thought about home and wondered what had been happening there while he and Pa were away.

He had not even started to get sleepy when he heard the six men who would be sleeping in the other beds coming up the stairs. The tipsy miner was singing, "Oh, my darling; oh, my darling; oh, my darling, Clementine."

"Hush, Charles! Yer gonna git yersef throwed outta hyere and ye'll have to sleep with the horses," one of the other men said.

The miner stopped singing and the men filed into the room. They noisily dropped their boots and britches and settled into their beds for the night. One of them blew out the candle, plunging the room into total darkness. Soon the sound of their snoring filled the room.

For a while, Clark stared into the darkness and wondered if he would ever be able to sleep. Drunken miners, Indian removals, tollgates and how to fly—so many things to ponder. But soon it all became a blur in his mind and his eyelids closed as he drifted into deep sleep.

5. Gainesville Adventure

The next day's travel from Auraria to Gainesville was the easiest part of their trip because the terrain was mostly level and the toll road was more heavily traveled between the two towns. When they reached the Chattahoochee River, they had to await their turn to pay passage onto a ferry for their horses and wagon to cross. Some of the other farmers who were waiting had cattle, sheep and hogs to take on the ferry. Using their staffs and herding dogs, they were able to drive their livestock onto the ferry without a great deal of difficulty. Clark was amazed at the skill of the dogs in keeping the animals moving in the right direction.

When their turn came, the horses resisted stepping onto the ferry, but Elisha calmed them and pulled the bridle firmly until they finally yielded and stepped aboard. As soon as they were on the ferry, the attendant waved to the man on the other side of the river. On cue, he began winding the strong hemp rope attached to the ferry, and they began moving across the stream.

Looking across the bow of the ferry to the other side of the river, Clark exclaimed, "Look, Pa! They have angled the rope to turn the ferry just enough for it to catch the river's flow to help push us along!"

Elisha was aware of this and felt very proud that his grandson had been observant enough to discover how it worked. "It's a pretty smart trick. Some of the loads that cross here are really heavy."

Upon reaching the other side, they disembarked and continued their trip, passing fields of corn and cotton that were far larger than any back home in Choestoe. It was a change of scenery for them because the farmers couldn't cultivate cotton in Choestoe's high elevation due to the growing season being too short.

By late afternoon, Elisha and Clark drove up to the home of Uncle Joe and Aunt Margaret. A pair of beagle hounds met them, barking at the horses and generally creating a lot of commotion. Joe and Margaret came out and upon recognizing Elisha, quieted the dogs and greeted him warmly.

Elisha said, "This is my grandson, Clark. Don't guess you folks have met him before."

"No, I don't believe we have," Margaret said. "Is he Sallie's boy?"

"Well, yes, he is," Elisha said. "But, you know, he has always lived with us and we feel like he is our son instead of our grandson."

Joe said, "Let me take your horses back to the barn, Elisha. Y'all go on in the house and let Margaret get you a glass of tea. I know you must be tuckered out from traveling so far."

As they went into the house, Clark was astonished at the sight of the large rooms and fine furniture. He had heard about Uncle Joe's wealth, but he had not imagined that his relatives lived in such luxury. As they sat in the living room, he could see their maid in the dining room, setting the table with fine china and placing silverware on large white napkins. A beautiful lace tablecloth was draped over the linen cloth that covered the cherry table, and it almost reached to the seats of the dozen chairs upholstered in dark blue velvet, matching the draperies that were hanging at the double windows. A patterned wool rug covered most of the floor.

The fireplace in the living room where they were sitting was faced with gray marble, as was the hearth. The chairs were upholstered in burgundy velvet with matching fringed draperies at the windows, and a fine rug of coordinating color was on the shiny oak floor. Clark glanced down to see if his brogans were shedding any dirt in this picture-perfect place. There were no homes like this in Choestoe, and he was thinking that it might be a good thing there wasn't. With the family going in and out all day across the dirt yard and fields, it wouldn't take long to make a mess of nice quarters like this.

Uncle Joe came in, hung his hat on the stand and sat down on the sofa beside Elisha. "Have you brought a lot of goods with you for trading?" he asked.

"We have our wagon completely filled," Elisha told him somewhat proudly. "I've brought corn, tobacco, flax, flaxseed and kegs of honey. How's the market for farmers right now?"

"Things are a-hoppin' in town. We've had a sight of folks move into Hall County in the past few years. It's got to where a feller can sell just about anything if it's in good condition."

"I'm sure glad to hear that. It's a long trip down here and I don't want to take any of my goods back with me. By the way, Elizabeth sent a list of cloth she wants me to get and a number of other things while I'm here. I'll need to find out from you where I can get everything."

"Sure, I'll be glad to ride in with you in the morning and show you around." Then turning to his wife, he said, "Margaret, if supper is ready, let's go in and get started. I know these fellers are hungry after their trip."

Margaret nodded, rose gracefully, and led them into the dining room. "With the children all gone, we don't have occasion to eat in the dining room very often anymore," she said. Waving to chairs nearest them, she said, "Elisha, you and Clark sit here."

Joe and Margaret took seats across the table from them, and the maid began serving the food, a platter of roast pork surrounded by roasted potatoes, green beans, creamed corn, steamed cabbage, and corn muffins.

Joe said grace and began passing the food around. Clark took generous portions of each dish as it came his way. The aroma told him that the cook was a good one, and he was going to take advantage of the opportunity and fill up on the good meal. It was a special treat after the plain rations they'd had on the road.

"Do you have a good school in Choestoe, Clark?" Joe asked.

"It's okay, I guess," Clark replied. "I have already finished the seven grades taught there. I borrow books from anybody who'll lend them to me. Living back in the mountains like we do, books are about the only way we have to learn new things."

"I want you to take a look at my library when you finish eating and see what books you want to take back with you. The only thing I ask is that you take good care of them and return them when you're finished. And take a look at the *Saturday Evening Posts* stacked in there. You might be interested in taking some of them with you. There's a lot going on in the world."

"I'm much obliged to you for that. I'll watch over your books like they're gold," Clark said.

The maid came in and asked, "Are you ready for some hot apple pie?"

"Yes, Hattie, go ahead and bring it in," Margaret said.

By the time Clark finished eating the delicious pie, he was starting to feel drowsy, but he was determined that he would not miss an opportunity to visit the library. As soon as the rest of them had finished eating and rose from the table, he said, "I'd like to see the library now if you don't mind."

Joe led the way to the library and upon entering the room Clark saw that one windowless wall had shelves from floor to ceiling, every one filled with books, just as Pa had told him. Joe said, "I have tried to arrange the books by subjects but sometimes that's not easy. Look around and see what you like. The old magazines are over there on that bottom shelf." He turned to Margaret and said, "Light the lamp for him, Margaret. It's about to get too dark in here for reading."

Margaret picked up the kerosene lamp on the desk and took it back to the kitchen to light the wick. Joe and Elisha settled into chairs by the desk and resumed their talk about business. Clark immediately became engrossed in searching through the books to see which ones he would borrow and was scarcely aware that anyone was in the room with him.

As Margaret returned with the lighted lamp, Clark exclaimed, "Hey, I found a biography of Benjamin Franklin!" He had become interested in Franklin during one of his classes and knew immediately that this was one of the books he wanted to borrow.

Clark continued pulling books from the shelves, glancing through them, then returning most to the shelf. He wasn't going to abuse his borrowing privilege by taking too many books. By the time the men stood up and announced it was bedtime, he had selected three more to take home: *The Pioneers* by James Fenimore Cooper, *The Life and Voyages of Christopher Columbus* by Washington Irving, and *The Birds of America* by John James Audubon. He carefully placed the four large books in a burlap bag, savoring the thought of having this treasure to take back home.

Margaret showed Elisha and Clark to the room where they would sleep. The bed had a full, fluffy feather mattress and pillows on it, and Clark had barely settled under the blanket before he was asleep.

Following a hearty breakfast served by Hattie the next morning, Elisha, Clark and Joe went to the barn to get the horses. Elisha hitched his horses to the wagon and Joe saddled his own horse to ride alongside them into town.

"We'll pass by the blacksmith shop first, Elisha, Joe said. "Do you need to get anything there?"

"I need to get an axe blade and rims for a couple of wagon wheels. Maybe I'll also get a plowshare if he has the kind I need, and might as well get some horseshoes too while I'm there," Elisha replied.

As they neared the blacksmith shop, Clark could hear the roar of the bellows fanning the burning coals and the clang of the smithy's heavy sledge as he beat the hot iron into shape. Smoke rolled from the furnace's hood, and sparks flew up in every direction from the yellow-hot iron bar as each blow struck it.

The burly man wielding the sledge stopped his work as they dismounted and walked toward them. "Howdy, Joe," he said, removing one glove and stretching out his hand to shake with him.

"You're always a busy man, Harvey," Joe said. "I want you to meet Elisha who's come to town to do some trading."

"Pleased to meet you, Elisha," Harvey said. "Can I help you with something?"

"Glad to make your acquaintance, Harvey. Let me take a look at some axe blades and wagon wheel rims. Also, I need to get shoes for my two horses over there. What about plowshares? I'll take a look at them if you have some," Elisha said.

As hot as it was in the shop at this early morning hour, Clark wondered how Harvey and the two men who worked for him could keep up the heavy work as the rising sun intensified the heat radiating from the furnace.

After Elisha got the items he needed, they decided to go to the general store where Elisha would trade out the goods he had brought from the farm for the rest of the supplies he was looking for. The store was located in the center of town with sheds in back where the horses could be tied and fed while the wagons were unloaded and the vendors transacted their business. Elisha pulled into one of the sheds and tied his horses. He and Clark then walked back around and entered the front door of the store.

Joe was leaning on the counter talking to the owner as they walked in. "Silas," he said, "I want you to meet Elisha Dyer. He's got a wagonload of goods for trading with you."

Silas nodded to Elisha and shook his hand. "Come," he said, "I'll get Henry to look at your produce and give you a price for it."

"Henry," he called as they went out the back door to the loading deck. "I need you over here to help a customer."

A tall muscular black man came from behind a stack of bags at the end of the deck. "Yes, sir. I can help him," he said. "Is that your wagon at the end, Mister?" he asked Elisha.

"Yes, that's it," Elisha said as he went to the wagon and dropped the tailgate.

Clark climbed into the wagon and as the men took bags and boxes to the deck, he kept moving the goods toward the tailgate for them. They soon had the wagon unloaded, and Henry began listing the amount of each type of merchandise. When he finished recording everything, he walked over to a ledger hanging from a chain on the wall and looked up the prices which he entered on his list.

"Now we'll take this inside for Mr. Silas to total," Henry said.

As Silas calculated the price of the merchandise, Clark walked through the store marveling at the variety of goods on the shelves, on the floor and hanging from the ceiling. Pungent aromas of coffee, tobacco and pickle barrel spices filled the air. A couple of men playing checkers on an upturned box in the only clear floor space in the store were calculating every move seriously, as if their lives depended on making the best possible maneuver. One of them kept scuffing his boots on the floor and stroking his scraggly beard nervously, periodically aiming a stream of tobacco juice toward a nearby spittoon. Clark smiled to himself at the idea of a simple game of checkers giving a grown man this kind of anxiety.

Moving to the other side of the room, Clark saw a woman looking through bolts of calico fabric on one of the shelves while her teenage daughters tried on one hat after the other, giggling quietly as they evaluated each other's choices. The mother finally called to them, "Come here, girls, and tell me if you like the cloth I picked out for your dresses. Your pa is going to be finished with his trading in a little while and we will have to go." They each grabbed a hat and came to her side.

As Clark moved down the aisle he came to a large display of pocketknives. He studied them carefully and wondered if he should ask Pa to get one for him. He knew if he got one, Jimmy would have to have one, too. Maybe the cost of two would be more than Pa thought he should spend. As he pondered what to do, Elisha came up behind him.

"They have some mighty good-looking knives here, don't they? I guess I can afford to get you and Jimmy one. I came out pretty good on my trading today," Elisha said.

"I would really like to have one. What do you think about this one?" he asked as he held up a Barlow.

Elisha took the knife and ran his finger down the blade and tested the grip of the handle. "Yes," he said. "I think this is a good one for you and Jimmy. Let's find the cloth Ma wanted and get finished here."

Clark took him to the shelf where he had seen the women looking at cloth, and they selected the colors they thought Elizabeth would like for herself and the girls.

"Pa, could we get hats for them, too? I saw some girls in here a while ago and they really liked the hats," Clark said.

"I think we can," Pa said.

They went to the shelf with hats and decided on white ones with blue ribbons for the girls and a green one with a black velvet ribbon for Elizabeth.

"I already have coffee, tea, spices, sugar, salt and kerosene set aside up front. Let's take these things and get some candy from the bin over there, then we'll be ready to settle up with Silas," said Pa.

It didn't take Silas long to get the cost of their purchases figured up, then he subtracted the total from the amount they had agreed upon for Elisha's goods. He paid Elisha for the difference and called to Henry to come help them load their purchases into their wagon.

As they drove toward Joe and Margaret's house, Clark was thinking how much he wanted to get back home. He had seen a lot of new things on the trip and in town today, but now he wanted to get back home and sleep in his own bed with his brothers. One more night in Gainesville, he thought, then we'll head for home.

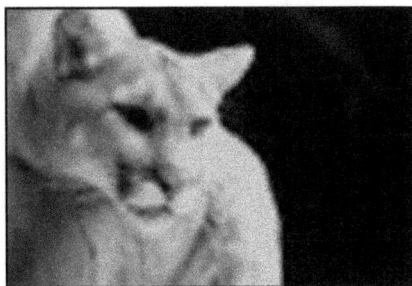

6. A Fright to Remember

As soon as they finished breakfast the next morning Elisha and Clark began preparing for their return trip home. Rain had started falling in the night and they sloshed through mud puddles as they loaded their personal luggage into the wagon. They spread canvas blankets over the bags and crates at the front and back of the wagon to keep the cargo dry.

When they had finished packing the wagon, they went inside to bid farewell to Joe and Margaret.

"I sure appreciate your hospitality and help, Joe," Elisha said. "I hope you and Margaret will come to the mountains before long and visit with us."

"It's been good having you with us, Elisha, as always," Joe said. "And good to meet you, Clark. We'd like to get up your way one of these days, but we might not get any farther than Habersham County. We are about to get too old now to cross the mountain and sleep in the woods at night."

"Uncle Joe, I want to thank you again for lending me the books. I know I will enjoy reading them," Clark said.

"Here's some food Hattie fixed for you," Margaret said, handing them a large basket. "It was good to see you both, and I want you to come back any time you can."

They climbed into the wagon, pulled their cloaks over their arms and legs and waved goodbye as the wet horses drew their load down the muddy road. Sheets of rain fell across the wagon, the horses' backs, and splattered on the trail before them. Elisha and Clark pulled their hats down to keep the rain out of their eyes and rode along without much conversation, each lost in his own thoughts.

As lunchtime approached and the rain showed no signs of abating, Elisha said, "We'll have to find some kind of shelter soon where we can feed ourselves and the horses. I think I remember an abandoned shack not too far up the road. We ought to be able to use it."

His memory had been accurate and it wasn't long before they came to the shack. The porch roof drooped on one side and weeds had grown tall around the little building. As Elisha pulled the horses off the road and into what had once been a yard, he reached for his sling blade.

"I don't want one of us or the horses getting a snakebite," he said as he climbed down and started cutting weeds and dragging them away from the porch.

Clark pulled the basket of food and a jug of water from under the canvas and carried them to the driest side of the porch. He loosened the horses' reins so they could eat weeds and grass while he and Elisha ate their lunch. He went to the door of the little house and looked inside. There was no furniture of any kind in the room, but straw piled against one wall was mashed down as though someone had recently slept on it.

"I guess we can eat standing up, can't we, Clark?" Elisha asked as he finished cutting weeds.

"That would sure suit me," Clark said. "I'm tired of sitting. My backside is sore after riding all morning."

They pulled back the cloth covering the basket and found big fluffy biscuits with pork tenderloin, baked sweet potatoes and slices of cheese.

"With a house servant who can cook like this, I bet Margaret has forgotten how to even make a biscuit," said Elisha. "Not that she really has time for cooking and housework any more. She keeps the records for all of Joe's businesses and tries to be at all the social events in town."

"I wouldn't want to live in town myself," Clark said. "I don't like to have a lot of things going on around me. Takes away my time for thinking and working on something new and different."

Looking at him with a teasing smile, Elisha said, "And what did you see in Gainesville, Clark, that you're anxious to get home and try to make?"

"It's not something I saw in Gainesville," Clark said. He hesitated a moment then added very seriously, "It was the birds we watched while we were resting on our way down here. I'm going to make something that will fly like a bird."

"Whoa!" Elisha said. "You may be making plans beyond your capabilities this time. I don't know anybody who can make a contraption that will fly."

"Maybe that's because nobody has ever tried to do it. I'm not thinking it would be easy or quick to figure out how to do it, but there has to be a way it can be done," Clark insisted.

"Well, it's something you can occupy yourself with when it's too wet to work outside, I guess," Elisha said, trying to humor him. "The youngsters will be happy with whatever you make."

Clark didn't say any more but he thought to himself that his first efforts at building a flying gadget might be only a toy for the kids, but the day would come when he would perfect a vessel that would really fly.

"Let's get rolling again," Elisha said as they finished their lunch. They threw on their cloaks, loaded the food basket, and climbed into the wagon. The drizzle persisted as they continued their drive. Heavy fog hung over the valley and blustery wind swept moisture across their faces and legs, penetrated their clothing, and left them feeling soggy all over. The physical discomfort and poor visibility caused the day to drag on interminably.

As night was drawing near, the Auraria stagecoach inn came into view. Clark and Elisha were relieved to get to lodging and quickly settled the horses for the night. The innkeeper recognized them as they entered and waved them toward the fireplace. "Sit a spell and dry out your clothes," he said. "It's been a dreary day for sure." Several other travelers were already seated by the fire and shifted to make room for them.

By the time food was served, they were dry and warm and ready for the hot meal before them. The group around the table this night was considerably quieter than the one they had eaten with on their trip down, but once again the conversation turned to the Indian situation.

"By Jove, they're going to have to force them to move out of here," said a hefty, red-faced man. "Besides the money the government gave them—five million dollars I think it was—they've also given them plows, axes, hoes, spinning wheels, and such like. Their leaders have signed a treaty with our government to give up the land anyhow. Why are they still staying here?"

"It'll be a sad day for lots of folks when they go," said the young man beside him. "Most of the Indians have been good neighbors to the white folks over the years, and there's been a good bit of intermarrying. Families will be torn apart when the removal comes. I know. My brother married an Indian girl. I don't think the government plans to make the ones married to whites leave, but each couple will have to choose which of the families they are going to stay with. I believe my brother will choose to move with the Indians because his wife would grieve herself to death if she had to part with her family. Of course, that means my brother will be separated from his family."

"I'm sorry for your brother's situation, but I think it's time for them to go," insisted the other man. "The Cherokee now have their own alphabet and newspaper. They even have their own laws and courts. It's just not workable for them to live among us."

"I guess that's the way the Cherokee chiefs see it, too. It's probably the reason they were willing to sign the treaty with our government," said the young man, as he excused himself from the table.

The rest of the group soon rose, bade each other goodnight and headed for their beds. Elisha and Clark climbed the narrow stairway to their attic dormitory as before, but this time their roommates were quiet and reserved. It didn't take long for the rhythm of the rain upon the roof to lull them all to sleep.

The next morning dawned clear and crisp. Elisha and Clark were grateful to have improved traveling conditions and looked forward to reaching Habersham by nightfall and seeing relatives again. Even the horses seemed happy to have sunshine on their backs, and they trotted along jauntily.

As twilight came, one small mountain remained to be crossed before they reached Amy and John's house. Weary with the day's travel, Elisha and Clark were riding quietly along the trail through the woods when suddenly a blood-curdling scream pierced the air. The startled horses jerked their heads, and Clark grabbed Elisha's arm. "Pa, what was that? I think a woman is screaming for help!"

Elisha reached beside him and lifted his gun. "No, son, I think it's a mountain lion. Hold the reins and keep the horses as steady as you can because they will try to bolt. We can't let that happen on this winding rutted trail or we'll lose all our load."

Clark took the reins and pulled back slightly to slow the horses. "Easy, boys," he said, trying to calm them. But at that moment another scream split the air, this time much closer judging from the sound. The horses whinnied and started to trot. Elisha called out, "Whoa" and Clark pulled hard on the reins.

Elisha jumped from the wagon and came to the side of the horses. "Steady, fellows," he said and stroked their necks. "Keep them moving as slowly as you can, Clark. I'll walk beside them and try to see where that hollering is coming from."

He didn't have to wait long before another shriek came from above them on the left side of the wagon. He leaped to the side of the trail and squinted towards the tree limbs. With wind swaying the trees and shadowy twilight blurring visibility, he struggled to see whether the panther was crouched on one of the limbs.

A sudden gust of wind bent the trees sideways and there was the big cat, ready to spring toward him. In a split second he raised his gun, lodged it against his shoulder and fired. The panther fell just two feet from him and struggled to get to its feet.

It took a moment for Elisha to recover from his fright and realize that the horses were galloping far up the trail, with the wagon bouncing wildly behind. He set off running after them as fast as he could but soon recognized this was a futile effort. Clark was the only one who could stop the horses now. Turning back, he went to see if the panther had been fatally injured by the shot. He found it to be still breathing, but there was little doubt that death was imminent. He wanted to wait and ensure its demise but decided he had better get up the road and find Clark and the horses.

Meantime, Clark was bouncing along the trail, struggling to get the horses stopped. He could hear the merchandise bumping loudly in the wagon and wondered what was happening to it. He pulled heavily on the reins but the horses seemed bent on putting distance between themselves and the screaming animal.

Finally, however, growing tired and not hearing further screams, the horses slowed their pace to a trot. Clark talked to them soothingly while pulling gently on the reins until they halted. He climbed down and pulled hay from the wagon to feed them. The tension had left him exhausted, and he breathed deeply as he stretched his arms and legs. He wondered whether he should tie the horses and go looking for Pa. If the panther happened to reappear while he was gone it might kill the horses. But what if Pa was hurt and needed his help? He decided the best course of action was to unhitch the horses and take them back with him.

Nightfall was coming fast as Clark went back down the mountain trail with the horses. He tried not to speculate about what could have happened to Pa. He had heard him fire the gun and knew he was a good marksman. Now all he could do was fervently pray that the shot had taken down the panther. He was grateful that the path was downhill, permitting him to make good time.

Elisha heard the clump of the horses' hoofs before Clark came into site. He yelled, "Hey! I'm down here." Clark broke into a run, pulling the horses after him. As he rounded the curve he saw Pa walking calmly up the hill with his musket over his shoulder. He ran to him and gave him a warm hug. "Pa, this is the scariest thing that's ever happened to me. Did you get hurt at all?"

"No, I saw that old scoundrel just in time to take aim before he jumped on my shoulders. What about you? Did you get hurt?"

"I'm fine, but I don't know what our merchandise will look like. It was really bumping hard as the horses ran away. I didn't take the cover off the wagon and check it before I headed back to see about you."

"Well, as long as we're both okay we won't worry a lot about the goods in the wagon. Let's get back up the trail and hitch the horses again so we can get to Amy's house before it's too dark to see anything."

The day that had started so pleasantly ended with a fright they would never forget.

7. Threat on the Horizon

Elisha and Clark were finally nearing Choestoe after spending more than a week making the trip to Gainesville. They had spent the night before last with relatives in Habersham County and their final night camped in the woods at Tesnatee Gap.

Twilight was approaching as they entered the valley and passed near the Indian Village beside Nottley River. The aroma of roasting venison and corncakes wafted through the air from the fires burning in front of the log cabins in the village. The Indians didn't cook inside their homes. They had acquired many customs of the white settlers, but cooking inside in fireplaces was not one of them.

"I'm hungry enough to eat a bear," Clark said, as he got a whiff of the food. "I hope Ma has supper ready when we get there."

"Me, too," Elisha said. "We'll have to take the wagon to the barn and leave it loaded until tomorrow. If Ma and the young-uns get started looking through the things we bought, Ma won't get supper on the table for an hour."

Clark laughed. "Jimmy and Lige will be fit to be tied if they have to wait until morning to see what we got them."

As they came up the homestretch of the wagon trail, the approaching hoofbeats alerted the dogs to their arrival, and the family came running out to meet them. Melinda and Matilda were bouncing up and down, with Elizabeth hanging onto a hand of each one to keep them from getting into the path of the horses. Jimmy and Lige were smiling from ear to ear as they leaped from the porch and ran alongside the wagon, anxious for Elisha to stop the horses.

Bluford and Cager stood in the doorway watching all of the commotion while doing their best to appear mature and above such emotional display, but their eyes sparkled with excitement at the sight of Elisha and Clark after their week long absence. Cager was holding Lump in his arms and he was squirming to get down and join his sisters in jumping up and down.

"Whoa," Elisha said as he brought the horses to a halt. He and Clark climbed down stiffly and he handed over the reins to Jimmy and Lige. "You can unhitch the wagon and leave it here till morning, Lige. Put the horses in the barn and give them some hay."

Melinda and Matilda pulled away from Elizabeth and threw their arms around Elisha's legs. He lifted them up, one in each arm. "Hey, my little angels! Did you miss your ol' Pa while he was gone?"

"You know we missed you, Pa," Melinda said, rubbing her cheek against his beard. "But we prayed for you every night."

"Well, that is why we made it back home safely. It was your prayers that did it." Looking affectionately at Elizabeth, he asked, "Did everything go okay here while we were gone, Lizzy?"

"We didn't have any big problems, nothing that the boys couldn't take care of," she said, coming to him and putting an arm around his waist. She reached her other arm out to Clark and hugged his shoulder. "Well, did you decide that you want to move to the big town after seeing it, Clark?"

"Ma, I don't want to ever live anywhere except right here in these mountains. It is really different in town and they have some nice things that we don't, but this place suits me a whole lot better."

Elizabeth laughed, "Come on in and wash up for supper. I know you're tired and hungry."

As they ate their meal, Elisha filled Elizabeth in on the news from the family members in Habersham County.

"They were really excited to see us and get the gifts you sent them," Elisha told her. "Amy was especially happy to get the blue cape you made her and said to tell you how much she loves you. They want to come over and visit us, but they have to look after John's daddy right now and can't get away to come. You'll have to let Bluford take you and Lump and the girls over and spend a couple of days with her soon."

"Well, I'll do that if you promise me you and the boys will do a good job of looking after yourselves while we're gone."

"Lizzie, we're not nearly as helpless around this house as you take us to be," Elisha said, giving her skirt a tug as she passed by him.

As soon as Elizabeth had the table cleared, Jimmy asked, "Pa, can we bring in the things you bought in Gainesville?"

"I thought we would leave them in the wagon until morning since it's so late," Elisha said. "But I guess everybody would rather open packages than sleep." Chuckling, he added, "Go ahead and get them. All of the bags you need to bring in tonight are behind the wagon seat."

Jimmy and Lige grabbed a kerosene lantern and lit it to take outside. Hurrying out the door, they jumped into the wagon and begin pulling out bags and placing them on the ground. As soon as they had everything unloaded, they began carrying the packages inside.

When they placed the first bags on the kitchen floor, the girls were all over them trying to get them open.

"Pa, don't let them open anything until we get it all inside," Jimmy said.

"Let's all lend a hand," Elisha said. "It's getting late and these little ones need to get in bed."

With extra help, it didn't take long to get the bags in the house, and finally all of the family could begin discovering what treasures were inside for them. Melinda and Matilda grabbed their new hats and put them on. Elisha reached over and put Elizabeth's hat on her head, and she tied the black velvet ribbon neatly under her chin, laughing at all the uproar.

Spying the bag that held the Barlow knives, Clark reached over, took one out and handed it over to Jimmy. He whistled and opened the blades, then reached for a piece of wood by the stove to test it.

"Don't cut your finger off, Jimmy. The blades are sharp," Clark said, feeling very proud to also have a Barlow of his very own now. "You know we will have to take care of these knives and not lose them. If we do, we'll pay a price for it. Pa won't buy us another one if he can't trust us to be responsible for keeping up with them."

"Let's make a holster for them to hang on our belts," Jimmy suggested. "We might lose them through a hole if we carry them in our pocket."

"Yeah, we can do that," Clark agreed.

Soon the young ones were tired out and Elizabeth led them off to bed. The girls got into their nightgowns and settled down quickly, but Lump was still too wound up to relax. She sat down and began rocking him, softly singing an old hymn. Before long it worked its magic, and she quietly carried him to his bed.

Returning to join Elisha and the boys in the kitchen, Elizabeth found them looking very grave as Bluford said to his father, "Pa, there has been a lot of talk around the settlement this week that President Jackson has issued an executive order requiring buyers of government lands to pay in gold or silver. They say the demand for gold and silver is becoming so great that many banks don't have enough of it to exchange for their notes. A lot of people think the banks are about to collapse."

Elisha said, "I'm not surprised to see it coming to this. Banks should never have been permitted to issue paper money with no commodity backing it. A lot of people have been speculating wildly in land. Public lands have a price fixed by law at $1.25 an acre and are open to any purchaser without limit as to area. There have been no price restraints like you'd have in free markets, and buyers have been able to pay with credit advanced by the banks. Of course, all of this created a bubble that is going to have to burst."

Clark couldn't see how all of this would have a bad effect on his family, but the look on the faces of Elisha and the older boys told him that there was a pending threat. He looked at Elizabeth and saw she too was aware of danger stalking them, though in a way that was incomprehensible to her as well.

"Mr. Henry said the sales of public lands that have been paid for in bank paper make up the bulk of the government deposits in the banks and that is how they have been able to extend such unreasonable credit to the land speculators," Bluford said.

"Well, they should have seen it coming. The country has had a period of grand economic growth and has been overconfident with lending and borrowing, turning a blind eye to the fact that this couldn't keep going," Elisha said. "I think we had better buy up supplies that we are not able to grow or make for ourselves and hunker down for some very rough days ahead."

"Elisha, do you think it will take a long time for the problem to be worked through?" asked Elizabeth.

"Don't you be worrying about it, Lizzy. I have kept out of debt and we've got the farm producing really good now. The people who are going to suffer are the ones who've put their investments in banks and in stocks of companies that have been speculating in land," Elisha said. "The people in the towns are going to feel the effects a whole lot more than the farmers will."

Clark watched the anxiety subside from the faces of the family as they trustingly received the comforting words of the family patriarch. As he so often did, Clark marveled at the vastness of Pa's wisdom. He promised himself that he would try to learn everything he could from Pa and be as much like him as possible.

Looking across the table at Clark, Bluford suddenly asked, "How would you like a job helping Mr. Henry down at the tannery?"

"I might," Clark responded. "If I can work for him when I'm not needed to help here at home. Do you know what he will pay?"

"He was paying Sam twenty-five cents a day when he worked there. Sam's family has moved to Kentucky and Mr. Henry needs to hire someone right away. He said he needs the person two days a week."

Clark looked at Elisha. "Do you think you can spare me to go to work for him, Pa?"

"Sure, if you want to do it," Elisha said. "It's dirty, smelly work, but Jake Henry is a decent taskmaster and I 'spect he would be glad to have you."

"I'll go see him tomorrow," Clark said. The thought of earning some money pleased him. He had gone to the tannery with Elisha several times, and he knew Mr. Henry dealt honestly with his customers. His sons were grown and owned large farms in Habersham County. He had one daughter still left at home but no one in the family to help in the business.

Elisha rose from his chair. "I'm heading to bed, fellows. It's going to feel good to rest my head on my own pillow again. And having Lizzy to keep my feet warm won't be bad either."

The boys smiled at their departing father and ambled toward their own room. Clark silently concurred with his Pa, it was certainly good to be home again.

8. Disquieting Discovery

Clark stepped out of the house and into the frosty air of the November morning. Days were growing short now and the sun wouldn't creep over the mountain range for another two hours. His walk to the tannery would be made in semi-darkness, but he didn't need a lantern to light the way. After more than a year of working in the village, he felt he could make the trip blindfolded.

As he strode along the path on this cold morning his mind was completely absorbed in rehashing the conversation he had had yesterday with Seth Harvey. Seth was two years older than Clark, and his family had moved from Habersham to Union County with the other families that came the same year as the Dyers. Since early childhood, Seth had been known for picking fights and taunting the other children in the neighborhood, especially the younger ones.

Seth had brought some hides to the tannery and while Mr. Henry was talking to another customer, he had sauntered over to Clark and cagily said, "You don't even know who your real Pa is, do you?"

"What business is it of yours?" Clark had retorted, instinctively clenching his fists. "Elisha is the only Pa I know and he's the best one anybody could have."

"Yeah, but now that you are nigh grown, I'd think you'd want to know who your real Pa is."

"And I suppose you think you know who it is."

"I don't know except what I've heard, that he was a German feller who lived pretty close to your grandpa in South Carolina. They say your grandpa never liked him and everybody was shocked that the feller had managed to get alone with Sallie behind his back."

Mr. Henry had returned at that point in their encounter and asked Seth to bring his hides over to a table across the tan yard to be counted. Seth had given Clark a taunting half smile as he left, seeming to know that he had planted a seed that would take root and grow in Clark's mind.

Well, he had guessed right, Clark thought, because he had been pondering the question a lot over the past couple of days, finally deciding he would bring up the subject with Sallie at the first opportunity. She and Eli usually came to visit them on Sunday afternoon, and he thought that would be the best time to pull her aside and see what he could learn. He was hoping he could open the subject without making her feel he was dissatisfied with the way he had been raised by his grandparents. He secretly preferred his grandmother's style of running her household over Sallie's. Ma appeared to be able to manage every-thing effortlessly and gave close attention to the needs of everyone in her large family. Sallie, on the other hand, had a tendency to flit from one task to another, and her children were sometimes forgotten as her interest was caught away by some event or conversation.

Clark couldn't imagine what kind of life he would have had if he had been brought up by someone other than Pa. Who else would take almost every opportunity that presented itself to teach him about farming, nature, people, building things, and enjoying life? Eli seemed to be a good father to Clark's half-siblings, but Clark felt sure they did not have the same quality of relationship that he had with Pa.

He searched his memory for everything he had been told or overheard about his out-of-wedlock birth. He knew that Sallie had continued living with his grandparents until she married Eli.

At that time he was only two years old. Even after their marriage, they had lived in the house with Elisha and Elizabeth until they all moved from Pendleton, South Carolina, to Habersham County, Georgia, when he was four. He had grown up calling her Sallie, just as the rest of the family did, and she was like a beloved older sister to him.

Ma and Pa had been his parents in every way all of his life, so much so that up to this time he had never given any thought to the events giving rise to his birth. But now that the question had been planted in his mind, he could barely wait to find out just how it had happened.

Following the big family meal the next Sunday, Clark walked across the porch to the swing where Sallie was sitting. "Why don't we walk over to James and Lucy's house and see how they're doing?" he asked her.

True to her nature, she brightened at the suggestion of going somewhere and immediately got to her feet. "I think that's a good idea, Clark," she said. "Let me go tell Ma we're going and tell her to keep an eye on the kids for me."

They had gone only a short distance down the road when Clark broached the subject. "Sallie, tell me about when I was born, about who my father is, and why you didn't marry him."

She was startled. "Clark, why are you asking about this now? That was a long time ago and I have put it all behind me."

"But think about if you were in my shoes, being a teenager now and not knowing who your father was. Wouldn't you want to find out?"

She walked along without saying anything for a few minutes. He knew if he kept quiet she would begin talking after she thought about the events that had taken place in her life at that time. He cast a sideways look at her walking along the mountain path and realized that she must have been a very pretty girl back in those days. She was still attractive. Her chestnut hair was always neatly arranged in a bun at the nape of her neck and her dresses fit well over her trim body. He thought of her as middle-aged, but she was only thirty-three years old.

"Clark, what I am going to tell you will make a lot more sense to you when you are older. You're going to wonder right now how I could have let it happen, but please realize that when it all came about I was just a year or two older than you are."

"His name was John Meyers and his father had a clock repair shop several miles from our house. He was a big muscular guy, five years older than I was. His family was already living in Pendleton when we moved there.

"He was a big flirt and always teased me when I was looking around the shop as Pa and Mr. Meyers discussed business. He would say things like 'You know you're the prettiest girl in Pendleton. Your shy little smile just makes my day. Look at you blushing! Now you really are a beauty.'

"Being young and naïve I was completely taken in by all of this. Pa noticed that I was always blushing and flustered after John had been talking to me and guessed that he was flirting with me. I suppose everyone in Pendleton knew John was flirtatious. Anyway, Pa stopped taking me with him when he went to the shop, and I'm sure it was to keep me away from him.

"One day I was on the way to Aunt Polly's house and was overtaken by John in his father's carriage. He stopped the horse, climbed down and took me by the hand. He said, 'I can't believe my good fortune in finding my pretty little sweetheart walking down the road today all by herself. Let me help you into the carriage and give you a ride.'

"I got in the carriage beside him but said very little as we rode along. I didn't really need to say much because he was a very big talker himself. When we got to the fork in the road that went to Aunt Polly's house, he turned in the other direction.

"'Wait, John,' I said. 'This is not the road going to Aunt Polly's house.'

"'I know,' he said. 'But I want to show you where I would like to build my house one day. It's a pretty place not far down this road.'

"We rode maybe a half mile and came to a little valley where he stopped the horse. He helped me down and led me over to a little spring branch.

"'This is the best water you've ever tasted,' he said. 'And the spring is big enough that I can build a spring house over it to keep milk and butter really cool.'

"I was so impressed with all of his big plans and flattered that he wanted to share them with me.

"We got back in the carriage and he returned to the road going to Aunt Polly's house. He stopped when we got to the path leading to her house and said, 'Sallie, I'd better say goodbye to you here. I hope I can see you again soon. I would come calling at your house, but I know your Pa doesn't care for me. Maybe I can win him over one day; I sure hope so.'

"I saw him several times after that—at a barn-raising for one of the neighbors, a corn-shucking party and a church picnic. He always made a beeline for me when he saw me and talked excitedly about the things he was doing. He never failed to tell me how pretty I was, and I fell completely in love with him.

"One day in late October I walked down to the village with some chestnuts I had gathered to barter for a pair of shoes. As I went by the clock shop, I was surprised to see a sign on the door "Closed." I asked at the shoe shop why they were closed and the owner said Mr. Meyers was moving to Mississippi along with several other family groups. Of course, I really hated to hear that because I assumed John would be moving along with his family.

"After I bought the shoes and began walking back home, I heard someone call my name and turned to find John running to catch up with me.

"'Sallie, I want to talk to you. Come back to my dad's shop and sit down for a few minutes and let me tell you what is happening.'

"We went back to the shop and sat down on the settee to talk. Although no one was there, John talked very quietly. He was somber, not at all his normal cheerful self. He said his father felt that he could get a lot more business in another town and that was why they were moving away next week.

"'Sallie, I haven't said this before, but I guess you know it anyhow. I love you and I don't want to leave you behind.'

"He put his arm around my shoulder and brought his face near mine. His blue eyes were filling with tears as he begged barely above a whisper, 'Please go with me. How can I go away without you, Sweetheart?'

"I tried to say that I couldn't just leave with him, barely knowing his family as I did, but he was kissing my forehead, my eyes, and my lips. Passion overcame us and we did what we should not have done.

"As we left his dad's shop that afternoon, he promised he would come back for me in a few months and we would be married. I trusted completely that he would do as he said.

"The weeks went by and it became obvious to me that I was expecting a baby. There had been no word from John since his family moved away. I knew I had to confess my situation to Ma and Pa, and I dreaded facing their disappointment. I dreaded facing what the future held for me. It seemed that of all the sins a girl could commit, the worst one in the eyes of the community was to have a baby out of wedlock.

"It was nearing Christmas and snow was on the ground. Ma was sitting by the fire knitting wool socks. I was sitting near her shelling corn to take to the mill. The younger kids were at school. I kept trying to get the words out but couldn't think of any way to break the news to her. Finally, I just blurted out, 'Ma, please don't hate me. I have to tell you something bad.'

"'Well, go ahead and tell me, Sallie. You have been moping around here for days looking like a sick puppy,' Ma said.

"'Ma, I'm going to have a baby—John Meyers' baby,' I said.

"I burst into tears and Ma was at my side in a flash. 'Honey, we will help you deal with this. You don't have to carry the burden by yourself,' she said. Then she asked if John knew about it.

"I told her he had promised to come back and marry me but that I had not heard anything from him. She said she would talk to Pa and they would get in touch with him. But as it turned out they didn't get in touch with him because when Ma told Pa what had happened he said he was not going to have any such fellow married to his daughter.

"I have no doubt that John would have come back and married me if he had known I was having his baby. But who knows if that would have worked out to be the best thing? I am so attached to my family and he was just as attached to his. I believe he would have been miserable living in Pendleton without them, just as I would have been miserable living in Mississippi without my family.

"I married Eli when you were only a year old, and Ma and Pa didn't want me to take you away from them. That is how they came to raise you. Eli and I lived just up the road, and I was able to be with you even as I began raising my other children. It actually worked out well for all of us.

"If some of the neighbors ever try to make you feel inferior because of the circumstances of your birth, don't let them do it. The mistake I made has been more than compensated for by the good lives we have lived since then. You are an exceptionally bright young man, and all of us are just as proud of you as we can be."

After she finished relating the long story to him, Sallie looked over at Clark to see how he had received it. He stopped walking and wrapped both arms around her.

"Thanks for that, Sallie. I'm glad to know the whole story. I would be interested to find out what happened to John, but maybe it's best to let that be a closed chapter. Likely, he also has a wife and family in Mississippi, or somewhere. It might unsettle him to discover that he has a son he never knew about."

That was the last time the subject was ever brought up between Clark and Sallie, but the history of his birth that was shared on the walk that day created a new bond between them that would last their lifetimes.

9. News of a Panic

A rriving home at sundown from a trip to town, Elisha came into the house wearing an extremely grim expression. He was carrying a newspaper in his hand, which he held up for the family to see the bold headlines: "Banks Collapse, Stock Markets Crash."

"This was bound to happen," he said dismally. "Now the country will fall into a terrible depression. It will take years for business to get back to normal."

Clark jumped from the chair where he had been whittling pegs for the table he was making, scattering wood chips across the floor as he ran to see what the newspaper had to say about the alarming event. He recalled the conversation a few months back between Pa and Bluford about the possibility that something like this was going to occur because of the wild speculation in land and the inflated loans the banks were making.

He took the newspaper from Elisha and read story after story about how the U.S. currency was depreciating, unemployment was skyrocketing, and food prices were escalating. A lot of people were blaming the economic collapse on former President Andrew Jackson's Specie Circular, which was an executive order Jackson had issued requiring federal land buyers to use gold or silver, rather than paper money or credit, to pay for the land.

"Pa," Clark said, "Here's an article about something else. It says President Van Buren is going to soon enforce the removal of all the Indian tribes from Georgia, Tennessee, Alabama and

South Carolina. It says he is making plans to have them rounded up and moved to Oklahoma. Do you think they will make the Cherokees living around here move, too?"

"I'm sure of it," Elisha said. "I hear they have already started building camps where they will hold them until they begin the trip West."

The family fell quiet as each of them absorbed the shattering news. Cager and Lige sat staring at their feet, wondering if it would affect their jobs at the Coosa Gold Mine. Jimmy sat with a puzzled look on his face, scratching the dog's ear and waiting for someone to say what all of this would mean for the family. Melinda and Matilda were too young to understand the problem but, feeling the anxiety of the others, they too were silent as they sat by the window embroidering. Even Lump, playing on the floor with his wooden carts, did so without his usual sound effects.

Elizabeth was the first to speak. She looked around the room at each of them. She had her hands on her hips and her green eyes were flashing defiantly.

"I am confident we have made adequate preparations for this time of trouble. I'm not going to be worried about it, not one bit. We don't owe the bank anything or anybody else. Sure, the price of cotton has gone sky high, but our family over in Habersham can supply us with all the cotton we need for our homespun goods, and we grow enough crops on this farm that we can barter for anything else we must have. We weren't relying on the banks or the government before this happened, and we're not relying on them now."

"Hear, hear," Elisha said, his blue eyes twinkling in amusement. "The decree has been given. Ma will thrash any ol' government that tries to threaten this family's well-being."

Then he turned serious. "What she says is true. I don't expect us to be very much inconvenienced by these things. But I don't want you all to be talking to friends and neighbors about the situation. There will be a lot of hot-headed people making accusations against the government and against owners of the banks. There will be fights and lawsuits. You have nothing to gain but trouble for yourself if you get drawn into it. Encourage everyone that all of us are going to survive the hard times just fine. A lot of people are going to need encouragement because there are some families that don't stand shoulder to shoulder like we do when trouble comes upon them."

Feeling better about where they stood concerning the economic panic, Clark's thoughts turned to the other item of bad news.

"What about the Indian removal, Pa?" he asked. "Is there anything we can do to keep the government from making the Indians go to Oklahoma when they don't want to go? The families living here are not going to want to go, and I would really miss Adahy if his family moved. He's been one of my best friends since we came to Choestoe."

"His father went with some other local Indian leaders to see Governor Gilmer at Milledgeville," Elisha said. "They petitioned him to let them stay on their native lands, but the Governor reminded them that the Indian members of the Treaty Party in 1835 had signed a legal document agreeing to the removal of all Cherokees to Oklahoma."

"What a terrible thing to do!" Clark said. "Why would they betray their own people like that?"

"I guess we will never know what they were thinking. Maybe they began to feel that trying to deal with the laws of both the United States and the Cherokee Nation was creating too much friction and confusion. Certainly, when we white people moved in and took over their tribal land and began to outnumber them, they must have felt that the only way they could recover their former way of life was to agree to the pressure that was being put upon them by our government to move to another territory where they would be with their own people and under their own laws."

"Matilda Owenby mentioned yesterday when she was here with Morena for their spinning and weaving lessons that the two Owenby boys who married Cherokee women are getting ready to move to North Carolina right away before the roundup begins. In fact, the whole Owenby family may eventually move out of Georgia to join their Indian relatives," Elizabeth said.

"I think that's the smart thing for mixed families to do. It will keep them close enough to visit each other now and then," Elisha said.

Then, frowning, he added, "I really don't feel good about the government's plan to move the families all the way across the country. I remember how the move we made from Habersham to here was quite hard on many of our people, as well as for the animals. Some of them wouldn't have survived if the move had been many more miles than it was."

"Adahy has been helping me decide which materials will work best in getting my flying model to operate," Clark said. "He's about the only person in the neighborhood who takes an interest in it. I wish he could stay here."

"Maybe you should let him take the contraption with him if he moves away, Clark. I don't believe you will ever get it to fly anyhow," Cager said.

Clark looked at him. His eyes narrowed and his lips tightened, but he decided not to respond to the remark. What good would it do? Cager had no vision of a person flying like a bird. In fact, he rarely even understood how to modify equipment around the house and farm to make them operate better. When Clark was altering something, Cager protested the entire time, complaining he wasn't doing it right and it wouldn't work properly if he fixed it that way. Many times even after he finished a project and demonstrated the improvement, Cager would insist that the other way was better. Clark wasn't sure if it was jealousy or just his preference for keeping everything the way it had always been. But whatever his reason, Clark wished he would keep his opinions to himself, especially about the flying machine. He knew it was going to operate perfectly when the kinks were worked out. He was going to surprise a lot of people the day he launched it into the air and it flew across the field.

Elisha said, "I can see that things are going to change all across the county because of the bank failures and the Indians being removed. It's unfortunate that two such major upheavals are happening at the same time."

He looked around at his family, knowing they relied on him to provide explanations that would help relieve their anxieties.

He continued, "You know, the Europeans were immigrating here many years before we came and have, for the most part, lived peaceably with the Indians. But that doesn't mean there haven't been some greedy men who cheated quite a few of the Indians, and there have been troublemakers from both sides that have caused discord between the races. The inter-marrying has created its own set of problems for both sides, too. It grieves me that the end result will be such hardship for the Indians. The ones who live here do not deserve to bear the consequences of deals made by their tribal leaders."

"I wonder if the Indians will be able to get fire water in Oklahoma," Lige interjected. "Seems like to me they are buying half of everything that's produced by the stills here in Union County."

"Oh, I think you can be assured there will be stills in Oklahoma," Elisha said. "But the Indians don't drink as much as you think they do. It just doesn't take much strong whiskey to get an Indian drunk, and they seem to get addicted to it a whole lot quicker than Caucasians do."

"Well, it's a shame our people got them to drinking the stuff," Elizabeth said. "I hope nobody will be making it in Oklahoma. I've seen too many of their women and children who've been hurt by drunk men. And they hurt themselves and each other, too. I'm sure the good Lord never intended for people to drink hard liquor. It ought not to be used for anything but medicine."

"Ma," Melinda said, coming to Elizabeth's side, "I want to make a rag doll for Tadewin to take with her. She loves playing with mine, and it will give her something to remember me by."

"That's kind of you to want to do that, Melinda. Of course, you can make a doll for her," Elizabeth said. "Why don't we make gifts for all of her family? I guess it will have to be small gifts because they will probably have to leave some of their things behind for lack of space in the wagons when they leave here."

"I'll make toys for the boys," Clark volunteered. Shooting a look at Cager, he added, "I'll make a flying vessel for Adahy like mine."

"I will knit a shawl for Habatu and one for each of her big girls," Elizabeth said.

"I would like to give Kashia one of my books to take with her, Ma," Matilda said. "Will you let me do that?"

"Yes, that will be a good gift for her, Matilda. I have seen how she beams at you when you let her look at one of your books," Elizabeth said.

Elisha listened to his family making their goodbye plans for the natives they had come to know and care for in the five years they had lived in Choestoe Valley. He thought it best not to make any further mention about the likely hardships these people were going to face on the journey that lay ahead of them. He would just join in making the farewell as pleasant as he could for everyone and pray for the best for them.

10. Interesting Visitor

S now had fallen in the night and was still drifting down as day broke across Choestoe Valley. The fresh white blanket that covered the fields gave a new and exotic look to the farm, Clark thought, as he gazed across the meadow toward the ice-fringed creek bank. Feeling that certain dank chill in the air and smelling the peculiar fragrance of the snow-covered earth brought to his mind the early childhood pleasure of running across the meadow to fall face-first into a snowdrift.

As he headed to the barn with Jimmy to milk the cows and feed the livestock, Clark strode along, breathing deeply of the chilled air and thinking that Ma would surely make snowcream if this kept up. It was a treat she enjoyed creating for the family, and it certainly was one that they looked forward to having.

"I guess Lump and the girls are going to expect us to take them sledding if the snow gets much deeper," Jimmy said to him.

"I'm sure they will," he replied. "That will give us a chance to see if the big sled we made is good enough to hitch to the horse and have him pull us through the snow."

Clark had another very special reason for being happy to see the inclement weather. It would free him from work so he would finally have a chance to steal away to a quiet corner and finish reading one of the books he had borrowed from Uncle Joe. He had started reading the *Autobiography of Benjamin Franklin* and was finding Franklin's quest to learn of better ways to make and do things so much like his own tendency. He had recently adopted Franklin's practice of recording every step he took in building or improving an item, then entering the results of whether it succeeded or, if it failed, why it did.

He had bought a record book in which he was keeping a careful account of the weight and measurements of each part of the airship model he was building. Several years had passed since he first got the idea in his head that a person ought to be able to build a machine that would fly like a bird, but his dream remained very much alive. Sometimes when an idea struck him at night of how some part of the machine should be designed, he would rise from bed, retrieve his record book from under the mattress, light a kerosene lamp and go to the kitchen to write down details of the idea that had come to him.

As he sat on the stool in the barn stall with his head resting against the cow's flank, both hands squeezing milk into the pail, his mind wasn't on the farm chores at all. His brain was busy concocting ways he could get materials for his flying model that would be lightweight, yet malleable and strong enough to permit him to launch, fly and land his machine without breaking it.

Jimmy entered the stall where he was milking and brought his thoughts back to earth with a thud. "I'm through feeding everything," he said. "Are you about finished?"

"Yeah, yeah, just getting the last of the milk right now," Clark replied.

He finished up quickly, and when they returned to the house, they found their three younger siblings up, dressed and waiting for breakfast. Having discovered that it was snowing, they were excited and raring to go outside and play.

As soon as they finished their breakfast, the whole group got on their gloves, coats and boots, and out the door they charged, ready for a new adventure. Lump, Melinda and Matilda started rolling snow to make a snowman. They had the first ball growing nearly large enough for the base when Jimmy and Clark came around the house with the horse and sled.

"Whoa," Jimmy said, stopping the horse beside them. "Who wants to go for a ride?"

"Me! Me! Me!" they all squealed and, leaving their barely commenced snowman behind, they scrambled aboard.

"We're going over to Eli and Sallie's house and pick up Andy, Bud and Tom to take them riding with us," Clark said.

The sled proved sturdy enough for skimming along the frozen wagon trail with its load of youngsters, but the snow was concealing rocks and potholes and that caused quite a bumpy ride as the skids kept hitting them. No one was complaining about that though. They were in high spirits and the bouncing seemed only to add to their joviality.

As they approached Eli and Sallie's house, they noticed an unfamiliar covered wagon sitting in the yard and wondered who would be traveling in this kind of weather. Hearing the commotion of the youngsters, Eli came out of the house to see who was there.

"Lord, have mercy!" he exclaimed. "What are you kids doing out here in this snow?"

"We're trying out our new sled and wanted to come and take Andy and Bud and Tom for a ride," Clark said.

"Well, come inside and sit by the fire awhile and get warmed up so you won't catch pneumonia. We have an interesting feller who stopped in on his way from Gainesville to Toccoa," Eli said.

They tramped inside and saw a brawny man with wild red hair and beard sitting by the fireplace with his stocking feet resting on the hearth, his knee-high boots propped against the chimney. His black felt hat and long wool coat hung on pegs at the door. His nose and cheeks were glowing crimson, having warmed up after being thoroughly chilled from the winter weather he had traveled through.

"Sampson, these are my relatives from up the creek. This is Clark, Jimmy, Melinda, Matilda and Lump," Eli said. "And this, children, is Mr. Sampson Picklesimer who has traveled all over the country."

Sampson smiled as he looked over the youngsters and said to Eli, "You folks sure raise some mighty fine-looking offspring in these mountains. They must be pretty tough, too, to be wandering around in this kind of weather."

"Mr. Picklesimer won't be able to cross the mountain with so much snow on the ground, so he'll have to wait it out here at least until tomorrow," Eli explained to them.

Sallie came in the back door with Andy, Bud and Tom trailing behind her.

"We have three buckets of snow on the porch and I'm going to heat some creamy milk with eggs and sorghum to make snow-cream for us," Sallie said.

She mixed her ingredients in an iron pot and placed it on the grate in the fireplace. She had a long spoon for stirring and she pulled a chair near the fire where she could reach the pot to easily swirl the liquid as it cooked.

Mr. Picklesimer kept them all entertained with his stories about the places he had been and the people he had met while they waited for the mixture in the pot to cook. He told them he had kept a journal of all of his travels and he was planning to write a book about it one day.

"About three years ago, I spent a few days with a fellow in Louisville, Kentucky, who built gliders," he said. "He tried to talk me into riding on one, but I've read too many stories about how people have broken their necks when wind currents caused them to lose control and they crashed. He has been reading everything he can get his hands on about different designs other guys have tried in an effort to solve the problem of navigation."

"What is the man's name?" Clark asked eagerly. "I am interested in building a flying machine myself."

"His name is Edgar Carmichael. Say, you're a mighty young fellow to be aspiring to build a flying machine. But I'm sure Edgar would be happy to talk to you about what he's doing. Maybe you could get up to his place and take a look at the gliders he's built. I'll get his address out of my suitcase so you can write him."

"Oh, I thank you for that. I would really like to go up and see the machines he's built, but if I can't do that, at least it would be good to hear about the methods he has tried and found to work or fail," Clark said.

Sallie pronounced her snow-cream mixture done and rose from her chair to take the pot outside for cooling before adding the snow to it. The younger children followed her to watch the process as each big spoon of snow added caused the blend to become thicker and thicker until finally Sallie could no longer stir it. She took the finished product inside and spooned it into bowls for everyone.

"Well, this is certainly a special treat," Mr. Picklesimer said. "I have never had snowcream before. Write down the recipe, Mrs. Townsend, and I will try it myself when I'm visiting up in the cold country. As I was growing up in Savannah, we didn't get enough snow to do this. By jingo, it is good!"

When all of them had finished eating their snowcream, Clark said, "Come on, Andy, Bud and Tom. I want to take you for a ride on my new sled before we have to leave to go back home. Melinda, Matilda and Lump can stay here 'til we get back."

The three Townsend boys climbed onto the sled with Clark and Jimmy, and then Clark turned the horse toward Trackrock Gap.

"We'd better not go too far," he said. "I don't want darkness to catch us in weather like this."

The wagon trail toward Trackrock Gap followed the edge of plowed fields until it reached a creek that had to be forded by horses, along with any wagons or sleds they might be pulling. A flat log had been placed over the creek a short distance up-stream to provide crossing for people. If it had been a summer day, the boys would have stayed on the sled and let the creek splash upon them as they crossed. But with the freezing weather, they were not willing to get wet. They jumped off the sled as they neared the creek, and walked along the bank to the foot log.

"Place your feet flat and take small steps," Clark called to them. "You will really be sorry if you slide off the log and fall in the creek."

Clark stopped the horse and climbed upon its back to ride through the creek so he could keep control during the crossing. Stopping on the other side, he dismounted and waited for the boys to cross on the foot log. He wondered if he should have turned around on the other side so as not to run the risk of someone falling in the creek on a day like this. But now that all of them were safely across, he would take them as far as Henry

Holiday's house and see if Henry had everything he needed to survive through the bad weather before he headed back home.

Jimmy and the Townsend boys were excited to hear that they were going to Henry's house. They were sure he would have either a new dog or a wild animal he had captured and tamed because that was what he was noted for doing. He had lost his wife last year and without her to restrain him, the dogs and other animals he had tamed were allowed the complete run of the residence.

When they neared Henry's house, a half-dozen barking dogs came running toward them. Clark pulled on the horse's reins and spoke soothingly to try to keep him from bolting in reaction to the commotion. As soon as he brought the horse to a halt, the boys leaped off the sled and waited for Henry to come out, which he did shortly, smoothing his unkempt hair and beard as he walked toward them.

"Well," Henry drawled, "I shore didn't expect anybody to come up here today. I hope everybody is doing okay."

"Yeah, we're all doing fine," Clark said. "We just wanted to try out our new sled in the snow and thought we'd check on you to see if you needed anything."

"That's mighty neighborly of you. I actually could use you fellers to bring my milk cow up here from the pasture. I was wondering how I could get down there to milk her and climb back up that slick hill carrying a pail of milk without falling down and breaking my neck."

Clark said, "Jimmy, you and Andy go down and get the cow for Mr. Holiday."

"Many thanks, boys," Henry said. Turning to the others he said, "Come in, come in. I want you to see what my pet raccoon can do!"

Bud and Tom looked at each other, grinning in anticipation of what Henry might be able to get the raccoon to do. He certainly had a way of charming animals.

The raccoon was sleeping in a chair by the fireplace when they came into the house. She didn't arouse until Henry snapped his fingers and called, "Lula Bell." Then she stood up on her hind legs and fixed her shiny bright eyes on him.

"Come over here and unlace my boots, Lula Bell."

She jumped down and came over to Henry and quickly untied his shoelace and began pulling the string through each of the eyelets. When she finished, Henry said, "Good girl!" He took a peanut from a can and tossed it to her. She caught it and deftly cracked the shell to get at the nuts inside.

"I have to hide the peanut can because she can open the lid and steal all of my peanuts," Henry said with a laugh.

Jimmy and Andy came in and were disappointed that they had missed the raccoon's shoelace trick, but Henry was happy to repeat it for them.

"Is there anything else you need done before we go?" Clark asked Henry.

"No, no. That's a heap big help to get the cow up here. Thank you, boys, for doing it. I want to give you a little something for your help."

Henry went to a wooden box under the window, opened the lid and pulled out two slingshots, which he brought to Jimmy and Andy. He pulled the rubber sling back on one of them and then released it quickly to show the boys how much force it had.

"You ought to be able to kill a rabbit or squirrel on your way back home," Henry said.

The boys thanked him, obviously very happy to get a slingshot of this quality.

They then said farewell to him and started back down the trail toward Sallie and Eli's house. Clark breathed a sigh of relief when they were safely on the other side of the creek. As they continued onward, the boys were busy testing their slingshots by taking aim at trees as they went past. Clark was thinking that the day was coming to a close and he hadn't been able to do any reading whatsoever. He consoled himself with anticipation of getting the address of the glider man from Mr. Picklesimer and sending a letter off to him to find out if he had any new flying principles he might be able to use in building his machine.

Arriving at Sallie and Eli's house, the boys were glad to get back inside and cozy up to the fireplace again.

"How bad is the weather getting out there?" Eli asked.

"It's definitely getting colder," Clark said. "I think I'd better get Lump and the girls headed on toward home right away."

As soon as the boys were warmed up, Clark told the younger ones to get into their coats, hats and gloves and give Sallie and Eli hugs so they could be on their way.

"Mr. Picklesimer, I would like to get the address of the Kentucky man from you, if you don't mind," he said.

"Oh, yes. Be glad to get it for you," Picklesimer said and went to his saddlebag and pulled out his journal. He wrote the name and address on a piece of paper and gave it to Clark.

"Thanks," Clark said. "I hope you will be able to continue your trip by tomorrow or the next day. It was nice meeting you."

"Nice to meet all of you nice people. If a man had to get stranded, this is certainly the place to do it," he laughed.

The youngsters climbed onto the sled and huddled together to help shield each other from the wind. As they started on the final trail toward home, they could feel the temperature dropping and the children were beginning to tire. Clark didn't dare force the horse to move faster for fear the sled would hit a bump and throw them all into the snow.

Another half hour of riding brought them in sight of home. Pulling up to the porch, they stepped off the sled stiffly and all of them except Clark went inside to gather around the fireplace and thaw out. Elizabeth was checking Lump and the girls to see if they were okay from the cold trip.

"You kids missed the snowcream," she said.

"We missed yours, but Sallie made some for us at her house," Melinda said. "It was good!"

"Good for Sallie," Elizabeth said.

"How did the sled do?" Elisha asked Clark as he came inside from putting the horse in the barn.

"Did fine," Clark said. "I think we got the skids smoothed off good and got them attached to the sled at about the right angle. We went from Sallie and Eli's house to Henry Holliday's to see if he was making out okay in the bad weather, so we had to take it across the creek. I guess we gave it a good workout. I'll examine it when the weather clears up."

"Sallie and Eli had a stranger at their house, a fellow named Sampson Picklesimer who got stranded on the icy trail," Clark told Elisha. "He was quite a talker, telling about his travels across the country. I was interested in the story he told about a man in Kentucky who builds gliders, and he gave me the fellow's address so I could write him."

"If he has half the interest in flying that you do, I know you will enjoy hearing from him," Elisha joked.

"I'll be very happy if I can learn something new from him. There's not anyone around here who even wants to talk about flying. All they can do is poke fun at me for trying to build something that will fly. But I'm going to succeed in doing it. They will see," Clark declared as he headed to the bedroom to put the piece of paper with the Kentucky man's address on it in the box along with all of his other notes on designing the machine.

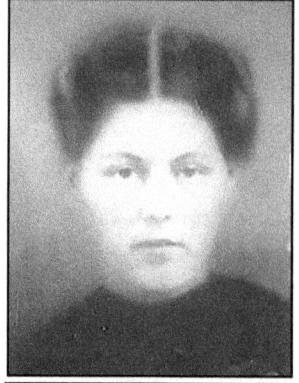

11. The Quiet Girl

Everybody in the community was talking about the festival that would take place in two weeks as they busily made preparations for it. The musicians and singers were honing their skills. The women were discussing what they would cook for the event. The girls were choosing which dress they would wear and whispering about which of the boys they hoped would bid on the box lunches they were planning to enter in the auction for raising money to pay the school teacher for the next year.

Clark, Cager, Jimmy, Melinda and Matilda were in the field harvesting the potato crop. As Clark walked behind the horse and plow, turning up the rows of potatoes, he was wondering if the Owenby family would be coming to the festival. He felt sure they would because all of them were actively involved in activities at the Choestoe Baptist Church where the event would take place. Morena Owenby and her mother had stopped coming to their house to take spinning and weaving lessons from Elizabeth about a year ago because they had become very proficient in both crafts and no longer needed lessons.

Clark wondered if he would have a chance to talk to Morena at the festival. She was always quiet when she came to their house, but with Ma and Mrs. Owenby talking and Matilda and Melinda chattering away, Morena hadn't had much opportunity to talk. He had never really had a conversation with her, and he was curious as to what subjects would interest her. He knew she helped a lot with home-schooling her siblings. There were eight of them younger than Morena, although the baby was now only a year old. Ma had told him that Morena really picked up quickly on how to spin and weave and did very nice work.

Clark wasn't sure about it, but he suspected that Ma was hoping Lige would get interested in Morena and perhaps she'd wind up being part of the family. They were the same age, nineteen, and three years older than Clark. The thought of those two courting amused him because Lige never paid any attention to Morena or any other girl. Clark figured if Lige ever got married it would be to the kind of girl who makes all of the advances, then drags the fellow to the justice of peace to tie the knot. Morena didn't fit that profile. She attended church with her family, and he was very sure anyone who got interested in her would have to do the pursuing and proposing. He liked that about her. Women could be spunky and still be appealing, but he didn't find anything attractive about a loud and aggressive female.

Another family that could be expected to attend the festival was the very large Ingram family from the Gaddistown community. They also were active in Choestoe Baptist Church and they had half a dozen very pretty teenage girls whom the Choestoe boys always wanted to see at community events. One of the Ingram girls usually managed to strike up a conversation with Jimmy, and as Clark thought about that, he had an idea for getting the two of them matched at the festival.

When they stopped their work and walked toward the house for lunch, Clark fell into step beside Jimmy.

"Are you going to bid on one of the box lunches at the festival this year?" he asked him.

"I might," Jimmy replied. Then he asked, "Are you?"

"I'm planning to, but I won't have much money for bidding so somebody will probably outbid me. But I don't care if that happens. At least it will cause some other fellow to put in more money for the teacher fund."

"The problem is that a person never knows which girl's box he's bidding on, so you could wind up having to eat with somebody you don't like," Jimmy said.

"The way to avoid getting into that fix," Clark explained, "is to get Melinda and Matilda to scout around and find out what the box belonging to someone you want to eat with looks like. All the girls tell each other what kind of box they are bringing and how it is decorated. Anyway, Melinda and Matilda can hardly wait until they are old enough to decorate a box and pack it with a lunch for the auction, so they will enjoy talking with the other girls and finding out what they did."

"It would be my luck that someone would guess that the box I kept bidding on belonged to a girl who was special to me and keep raising the price until I couldn't afford to get it," Jimmy said.

"That happens," Clark agreed. "But at least the girl will suspect that you knew it was hers from your continued bidding and be pleased that you wanted to eat with her."

The day finally arrived and it was blessed with azure skies and a soft breeze. Horses and wagons surrounded the church grounds, and the air was abuzz with talking and laughing. Everybody seemed to be in a jolly spirit as they arrived for the long-awaited festival. Box lunches for the auction were stacked on a makeshift table set up under a big sycamore tree.

"Let's get started, folks," Buck Collins yelled out. "I'm glad to see all of you-uns out this morning. We're going to begin with an old hymn, *Amazing Grace*." Motioning to the men with stringed instruments who were standing behind the table, he said, "Give me one sharp, boys."

They complied and the music began. The crowd sang out with gusto as Buck waved his arms in rhythm. After they had sung all four verses of the song, Buck asked for special requests from the crowd.

"Rock of Ages," someone called out, and before long the music and singing was again echoing off the mountainsides in beautiful harmony.

After the last hymn was finished, Buck said, "As I look over here at the table with all the box lunches, I think we had better begin the auction. Otherwise, it will be way past lunchtime before we get through taking the bids and finding out the winners."

"Here's how it works," he explained. "When the winner is announced, he should come up to the table and pay the amount of his bid to Sam over there. Then he should step over here to me and I will give him his box with the name of the lady who will share the lunch with him. From there, fellows, you're on your own."

Buck reached over and picked up a square box that was tied with a pink ribbon. He put the box in the middle of his right hand and moved it slowly up and down.

"I can tell there's fried chicken in here," he said with a big smile. "Now who wants to begin the bidding?"

"I bid a dollar," said Zeke Souther.

"One dollar," Buck called out in his best auctioneering voice. "Who'll raise it to one and a quarter?"

"I bid one and a quarter," responded John Owen. The eyes of the crowd turned to Zeke to see if he would continue bidding.

"I've got one and a quarter," called Buck. "Who'll give me one and a half?"

"One and a half," Zeke said boldly. Now the eyes turned to John.

"I've got one and a half," Buck roared. "Who'll make it one and three-quarters?" Eyes went back to John, but this time he remained silent.

"Do I hear one and three-quarters for this nice box?" Buck asked. Silence. "Then, going once, going twice and sold to Zeke for one fifty."

Zeke went to the table and handed Sam a dollar and fifty cents, then he went to Buck and took the box from him. Buck leaned over near Zeke and said the girl's name, but Clark and Jimmy couldn't make out what he said. They watched to see which way

Zeke headed, but he went back to the edge of the crowd and didn't look around for his lunch mate right away.

Buck picked up another box from the table. This one had a navy blue ribbon with polka dots. Melinda said, "I know whose that one is."

"Whose?" asked Jimmy.

"It's Morena Owenby's. I saw her put it on the table. I noticed that the ribbon on the box is made from the same cloth that Morena's mama's dress is made from."

Buck rubbed the polka dot ribbon and sniffed the box. "Woooee! Some girl has put pork tenderloin and apple pie in this one. It really smells good! Let's get started with a one dollar bid. Who'll give me one?" he called.

"I bid one," said a strong voice across the courtyard.

"Who'll make it one and a quarter?" asked Buck.

"I bid one and a quarter," said Clark to the astonishment of Jimmy, Melinda and Matilda.

"She's older than you," Melinda said. "Let Cager bid on her box."

"I don't have to *let* Cager bid on it. He can do whatever he wants to about bidding," Clark retorted.

"I have one and a quarter. Who'll give me one fifty," called Buck.

"I bid one fifty," came the voice from across the courtyard.

Now everyone was looking at Clark to see if he would keep bidding.

"I have one fifty. Who'll give me one seventy-five?" Buck asked, looking toward Clark, along with nearly everyone else.

Clark hesitated for only a moment, then said, "I bid one seventy-five."

"I've got one seventy-five. Who'll raise that to two dollars?" Buck asked. There was no answer from the other bidder, and Buck asked again, "Does anyone bid two dollars?" Silence again. "Going once, going twice, going three times, and sold to Clark Dyer."

Clark dug into his pocket and went to Sam with his money. Sam handed him the box and said, "This box was given by Morena Owenby."

Clark glanced her way and she smiled slightly. He made his way through the crowd to her side. "Hey," he said. "Buck told me this pretty box is yours. I hope you don't mind having lunch with me."

"Not at all," Morena said quietly. "There's a big rock over there we can sit on while we eat."

"That suits me fine," Clark said, "but first I want to play a joke on my brother, Jimmy. If you'd like you can go on over and claim a spot for us on the rock while I go back to talk to him."

"Okay," she said.

Clark worked his way quickly back to his brother and sisters. He leaned over and whispered to Melinda, "Which box belongs to Eliza Ingram?"

"It's wrapped in striped cloth and tied with a green ribbon," she said. "I think Buck will be putting it up for bid next."

She was right. As Buck handed off the box he had in his hand, he reached for the striped container and began his auctioneering prater, "Look at this pretty box. I bet some fine young lady has filled it up with ham and green beans and some really good biscuits. Who'll start the bid with one dollar?"

Clark stepped over to Jimmy. "I think you ought to try to get this one," he said.

"Why should I get that one?" Jimmy asked.

"Because it probably belongs to Sarah Jane Housley."

Jimmy blushed slightly, then looked over at the young man who was saying, "I bid one dollar."

"I've got one dollar. Who'll give me one and a quarter?" Buck asked.

"I bid one and a quarter," Jimmy said, turning a deeper red as the crowd turned interested looks toward him.

"I've got one and a quarter. Who'll give me one and a half?" Buck called.

The young man who had made the first bid didn't offer another bid and after a moment Buck said, "Do I hear a bid for one and half?" With no answer, he said, "Going once for one and a quarter, going twice . . . and sold to Jimmy Dyer."

Jimmy went forward and paid his bid and when Buck handed him the box he told him quietly, "The box belongs to Eliza Ingram."

Jimmy's mouth fell open and his eyes widened. He had tried to stay clear of Eliza and now he was going to have to eat lunch with her. Suddenly he realized that Clark had pulled a prank on him, but when he looked back to confront him about it, Clark was halfway across the yard, heading toward Morena. He looked over the crowd toward Eliza, and she gave him a wide smile. No need to fight it, he thought; might just as well go on over and join her. He would settle the score with Clark later.

Meantime, Clark reached the rock where Morena was waiting. She handed him her box and asked, "Do you want to open it and see what we're having for lunch?"

"I'm sure it will be very good," he said, as he untied the ribbon. "Thank you for bringing it to help in raising money for the teacher's salary."

"It looks like everyone's having fun, doesn't it?" she asked.

"Yeah, I believe they are," he said. "I hope Jimmy will get over being aggravated at me for making him think he was getting somebody else's box just now. Eliza has been trying to get him to talk to her for months, but he keeps sidestepping her. I think he will really like her when he gets to know her so I tricked him into bidding on her box."

"Well, I wouldn't have thought about you being a matchmaker, Clark," Morena said with a laugh, then added, "But I have noticed that you are much more observant than most fellows your age."

"Oh, you have?" he asked, surprised at her remark. "Now, that makes me think if you were watchful enough to notice that trait in me, it's likely I'm not the only observant person sitting on this rock."

They both laughed. Clark was glad he had bid on her box. Clearly, she was going to be a lot more enjoyable to talk to than nearly any other girl at the festival. And he was very pleased that she was not being condescending toward him because of his being three years younger.

"That was my sister Barbara's box that was auctioned off first, the one that Zeke Souther got," Morena said. "I don't suppose you mind if they join us," she looked at him inquiringly.

He really did mind. Zeke took every opportunity to poke fun at him about his idea of a flying machine, and Clark made a point of avoiding him as much as possible. But he could see that Morena wanted her sister to eat alongside them, so he said, "No, I don't mind. Motion for her to come on over and join us."

Barbara saw Morena waving to her and came across the yard with her long skirt and light brown hair blowing in the breeze. She was sixteen and, unlike her older sister, quite a flirt with the boys. Though Morena was the prettier of the two, when Barbara was around she was usually overlooked because of her quiet demeanor.

"Zeke has my box," Barbara said as she approached them. "I think he is teasing me by not looking me up for lunch."

"Ignore him," Morena counseled. "He'll come over when he sees that the delay isn't bothering you. I will go ahead and open my box and you can eat with Clark and me until he makes his way over."

Clark laughed when he saw what was in Morena's box. "Buck has a very good nose," he said. "You really did pack pork tender-loin and apple pie. My! It sure looks good."

As soon as they began eating, Zeke hurried across to them, tapped Barbara on the shoulder and said, "Hey, you're supposed to be eating with me."

"You're acting like you want to have it all for yourself, so I'll just eat with my sister," she said haughtily.

"Oh, don't be mad at me. I just wanted to have a little fun with you," Zeke said. "Come sit here beside me and open the box and we'll eat together."

Barbara turned around, looked into his imploring blue eyes for a moment and said, "Oh, well. I guess it is part of the bargain for a girl to eat with the guy who's the winning bidder on her box."

After a few minutes of conversation among them, Zeke turned to Clark and asked, "Have you got your machine ready to fly?"

"I have some more work to do on it yet," Clark said.

"Morena," Zeke said with a laugh, "Did you know that you're eating with a fellow who's going to fly like a bird?"

"Really, Clark?" she asked, "Are you making a contraption that will fly?"

Clark struggled to keep his composure, not wanting Morena to see how irritated he was with Zeke. "Yes, I am working on a model that I hope one of these days I will be able to set to flight."

"How interesting," Morena said. "I would like to see it. Will it be like a kite?"

"In some ways it will," Clark said. "But it won't be tethered to a string like a kite. I'm trying out different ways that it can be guided and powered to stay in the air."

"Well, when you get one ready to fly, let me know. I want to see how big a splatter you make when the thing crashes," laughed Zeke.

Clark didn't respond to the remark and changed the subject, asking Morena what kind of apples she used in the pie she brought.

Throughout the rest of their lunch the conversation was light-hearted and pleasant but, as they finished eating, Morena's brother Jonathan walked over and cast a solemn mood upon them with a grave announcement.

"Seth Adams says the government troops are coming our way to round up all the Indians tomorrow. They will put them in stockades until the removal to the West begins. Uncle Jim and his family have already packed up and are on their way to North Carolina. Uncle Hubert and his family will be leaving in a little while and will follow them out of Georgia. If you are acquainted with any mixed families that don't know about this, get word to them as soon as you can that if they are found here when the troops arrive they will be taken immediately to the stockades."

Tears filled Morena's eyes, as she said, "We will be losing some good friends as they are sent to the West. I pray God's protection for them on the long trip. Even for Uncle Jim and Uncle Hubert and their families it will be a hard trip to North Carolina."

As the news spread through the crowd, they began loading up their wagons and starting home, the happy atmosphere gone from the gathering.

Clark put his hand on Morena's arm. "I really enjoyed having lunch with you, Morena, and I'm sorry the day had to end like this."

"Me, too," she said. "My family may decide to move to North Carolina to join the ones who are having to go now. I will send word to you if we are going to move away."

"Yes, please do," Clark said. "I wish all of you well."

He turned and walked away with a heavy heart. After having lunch with Morena today he was even more impressed with her. Some of the things she had said gave him an anticipation that he would get to see a lot more of her in the coming days.

Oh, how he hoped her family would not be moving to North Carolina!

12. Sad Farewell

C lark and Jimmy were seated on a bench in front of the Choestoe General Store. The scene passing before them was the most distressing of their short lives as they watched a caravan of Indians move along the road. The natives were entering the village from each of the trails coming down the mountains on the south side of the Blue Ridge. They were being led by Federal soldiers who had rounded up a surprising number of the natives from the settlements along the Logan Turnpike and the Choestoe, Tesnatee, Trackrock and Enotah Trails.

These groups were being joined by others that soldiers had rounded up earlier, and now the caravan had grown quite large. Horses and cows tramped along with the displaced families.

The Indian children laughed happily as they skipped along-side the wagons, playing tag with each other in the early June sunshine. But the quiet adults walked with bowed heads, stooped shoulders and a deep shadow of sadness on their faces. They had finally been forced to leave their ancestral homes. Their freedom now gone, they traveled toward the stockade where they would be housed until their final trip began for Indian Territory west of the Mississippi, which would be their permanent home.

"I'm glad Adahy's family moved away before the soldiers came and captured them," Clark said. "They've been gone three months, but I don't think they have reached the Western Territory yet because Adahy promised to write me when they got there."

"Travel will be slow with the young children and animals," Jimmy agreed. "I miss them. I hope they're okay."

"They're a hardy bunch. If I had to make a trip like that, they would be the family I would choose to go with."

Suddenly, one of the soldiers stepped out of the group and came over to the bench where the young men sat.

"If you fellers would like a job, we're looking for a few good men to help with hauling supplies to Chastain's Station. That's the stockade at Blue Ridge."

Clark shot a questioning look toward Jimmy and he nodded his consent.

"We can help you for a week, but we'll have to return to our other jobs after that," Clark told the soldier.

"We can use your help even for a short time. Report to Sergeant Adderholt over there to get your assignment," he said, pointing to a tall, skinny soldier bringing up the rear of one group of the Indians.

They walked over to the sergeant and fell in step beside him. His uniform was rumpled and he wore his cap down nearly over his eyebrows. His musket, hanging across his shoulder, jostled with every step.

"Sir, the soldier told us you could use some help. We can work for you for a week," Clark told him.

Adderholt looked the young men over disdainfully. "Well, yer not the big burly men I need. I want somebody who can work for two weeks, not just one, but since yer my only choice, I'll take what I can get." Motioning toward the group to his rear, he growled. "Fall in with us now. We're gonna pick up supplies when we get west of Blairsville."

"Give me a minute to leave a note with the storekeeper to let our family know where we are," Jimmy said, as he hastily headed for the store.

When he finished writing the note and came back outside, the group was already out of sight. He broke into a trot and caught up with them about a quarter of a mile down the trail.

"Whew! It's going to be a warm day," he remarked to Clark as he joined him, wiping perspiration from his forehead. "If I had known this was what I'd be doing today, I would've put on a thinner shirt."

"You'll likely be happy to have that one when the sun goes down this evening. We'll probably have to sleep on the ground."

The sun was directly overhead when word was sent down the caravan line that everyone would be taking a rest stop in a short while. Soon the people and animals were directed to a clearing beside a stream where the bank was low enough for the horses and cows to drink. Enough food had been prepared at breakfast that morning to provide a cold meal for everyone.

Sergeant Adderholt handed tin plates to Clark and Jimmy. "You can go up to the lead wagon and get whatever the cook has today."

The cook was in Army uniform, standing behind pots that were sitting on upturned boxes. Clark and Jimmy held out their plates and he filled them with venison stew, beans and corn cakes. He waved his hand toward some tin cups stacked beside an oak bucket. "Git yerselves somethin' to drank."

They got water from the bucket and walked over to a large hickory tree that provided a good shade. As they sat, leaning against the tree, eating and looking at the Indian families, they were amazed at the complacency exhibited by them.

"But what choice do they have?" Clark mused. "If they resist, the soldiers will bind their arms behind them and separate them from their families. They will have to bring up the rear of the caravan and be last to get anything."

"The soldiers must have made the marching orders clear to them when they were rounded up since they are behaving so well," Jimmy said.

"Time to move out," the wagon master yelled. "You have about five minutes to wash your eating utensils in the stream and pack up."

Clark and Jimmy washed their plates, cups and forks in the creek and took them to the wagon up front. Sergeant Adderholt was waiting for them there.

"We'll be stopping about five miles ahead at Sam Pickens' warehouse to pick up two wagonloads of supplies. He will furnish a mule to pull each wagon. We will camp for the night in Sam's pasture, and travel on to Chastain's Station tomorrow. Hopefully, we will reach there by nightfall. You boys will be in charge of the mules and wagons. You will help unload the supplies the next morning and help settle the Indians and their animals at the stockade over the following couple of days. Then you will bring the mules and wagons back to Sam Pickens as you return home to Choestoe."

That sounded like light duty to the young men, and they were glad they had signed on for the job. The Army paid considerably better wages than the farmers or merchants in Choestoe, and the project had an additional bonus for them—an opportunity to see what a stockade was really like.

"Do you think there will be buildings with bars over the windows like the county jail?" Jimmy asked.

"Probably not," said Clark. "With this many Indians, I don't think there will be enough buildings to house them all. Most likely there will be one for holding troublemakers, and the rest will be in tents with armed guards keeping watch."

The caravan made its way through Blairsville and headed west as the sun was starting to dip towards the distant mountain range. The steps of some of the older Indians were beginning to slacken despite the wagon trail being smoother and the terrain level. As they gradually fell behind the main group, Adderholt marched over to them and barked an order: "Pick it up! It will be rough traveling fer ye if darkness falls before we reach our campsite."

The weary group struggled to walk faster. They spoke to one another in the Cherokee language, and from the tone of their voices Clark thought it sounded as if they were trying to give encouragement to each other. But their resentment toward Adderholt was noticeable and they shot dark glances his way.

Another couple of hours travel brought the group to Sam Pickens' pasture as the last light was disappearing on the horizon. The soldiers brought the wagons into a circle and herded the Indians and livestock into the center.

"Get food for the livestock from the haystack over there, and let the Indians get some for making their beds, too," Adderholt said.

Clark and Jimmy moved quickly to get hay moved inside the camp circle. It would soon be too dark to see what they were doing. By the time they finished, the only available light came from a few lanterns the soldiers had lit and hung on the wagon frames.

The Indians stretched out upon the ground in family groups, covering themselves with their bright blankets. A light breeze had arisen and without the warmth of sunlight or heat they generated themselves from walking, the cold ground was making an uncomfortable bed for Clark and Jimmy. They began to scrounge around for enough hay to make a pad they could lie on. They pulled some hay from the mules' feeding stacks to cover themselves, then lay down back to back.

"You're right about my shirt not being too heavy after all," Jimmy said. "I would be happy to have a jacket with me now."

"I wish we'd known when we left the house this morning that we'd be sleeping on the ground tonight. We could have prepared for it like the Indians did," Clark said.

Sleep came to the young men quickly despite the chill of the night, and they were startled awake the next morning by a poke from Adderholt's gun barrel against their legs.

"Come on!" he barked. "We're going over to Sam's barn to get the wagons and hitch up the mules. The cook will have breakfast ready when we get back. You fellers can bring up the rear of the caravan with the supply wagons when we hit the road again."

Clark and Jimmy stood up, brushing hay from their clothes and hair. Heavy dew had fallen in the night and their hair felt damp, but the covering of hay had kept their clothes dry. As they walked rapidly toward Sam's barn, they began to warm up again and became excited about the day that lay ahead.

They found the wagons packed with supplies from top to bottom. There were slabs of pork, barrels of cornmeal and dried beans, feed for the livestock and medical supplies. As they began harnessing the mules, Sam Pickens came into the barn.

"Morning, fellers," he greeted them. "I'm letting you use my two best mules. This one's Peanut and this one's Hank. Just a word of advice, you'll get better obedience from them with an apple than you will with a stick. I don't know if you've worked much with mules. They're good animals but they can sullen up on you if you make them mad."

Clark rubbed Peanut behind his ear and the relaxed drop of the mule's head gave Clark the assurance he needed that they were going to get along just fine. But when he laid his hand on Hank's shoulder, Hank craned his neck toward him and flared his nostrils as if to warn him to keep his hands off.

"Got any suggestions for handling this one?" Clark asked Sam.

Sam grinned. "Give him as much free reign as you can, but if he gets stubborn and won't go your way, crack him on the head with a pitchfork handle. It's the best way to get his attention."

Sergeant Adderholt said, "We've got to get started, Sam. Thanks for getting the supplies loaded up for us."

"Glad to have the business, Sergeant. I'll look for the mules and wagons back by the weekend."

"We'll have them here, won't we, fellers?" Adderholt looked at Clark and Jimmy.

"Yes, sir," they replied together.

After they finished breakfast, the soldiers led the caravan out of the pasture, onto the well-worn trail, and then headed west for Chastain's Station. Clark and Jimmy brought up the rear with the supply wagons. The mules seemed well trained, pulling their heavy loads and following the young men without resistance.

As the day advanced the sun began to bear down upon the group of travelers. All of them showed relief when word came down the line that they would be stopping shortly for a lunch break. They welcomed the cold stream running through the cove where the caravan stopped, and they eagerly quenched their thirst and cooled their faces before forming a line to get their food.

Clark and Jimmy led the mules to the stream for a drink before hitching them again to the wagons. By the time they got to the lunch line, they were at the very end. As the line advanced toward the serving table, Clark noticed one of the young Indian men kept looking their way. He nodded to him, not speaking because he didn't know whether he understood English.

Finally, the young man stepped back and asked in English, "You know Adahy?"

Surprised, Clark answered, "Yes, we are friends."

"I'm Muata, Adahy's cousin. I've heard him speak of you. My family should have left with Adahy and his family. Some of my people are not strong enough to make this journey under these conditions. My sister will have a baby soon and she should not be walking so many miles every day."

Clark reached for Muata's hand and shook it. "I'm happy to meet you, Muata. I will help you in any way I can while we are here. Do you know where Adahy is now?"

"Yes. His family is living beside the Great River to take time for gathering food and repairing wagons before traveling on towards the Indian Territory far west of their location. They have lost five members from families traveling with them, including Adahy's mother. I fear it will be much worse for us."

"I pray it won't be, Muata. May the Great Spirit go with you."

Muata shook his hand and moved back to take his place in line with his family. Clark saw that his sister was very large with child and knew that Muata had reason to be concerned about her welfare.

Clark and Jimmy devoured their food, seeing that many of the ones who were first in line had already finished and knowing that the wagon master would be calling for the caravan to get back on the trail again very soon. They had guessed right. His messenger arrived in a few minutes with word that it was time to move out.

After about an hour of travel, the trail became quite steep. The mules were straining to keep the wagons rolling up the precipitous terrain. Suddenly, Hank balked at pulling the heavy wagon and came to a halt. Clark realized that getting the wagon rolling again on an incline this steep would not be possible.

"Jimmy, keep Peanut moving if you can. I'll have to let this wagon roll back down to a place with less of a slope before I can get it going again. And that will only be possible if I can coax Hank into pulling again."

Clark unhooked Hank's harness and moved him to the back of the wagon. He pulled a log to the front of the wagon and secured it with a rope so it would drag and keep the wagon from gaining too much speed. Taking Hank's bridle he gave a gentle tug and urged, "Come on, Hank. Let's get this wagon rolling."

Hank pulled and the wagon began rolling downhill. Clark walked beside him urging him to go easy. Upon reaching the bottom of hill, Clark let the wagon continue up the next incline, and then stopped to move Hank to the front of the wagon again and untie the dragging log.

He patted Hank's shoulder and got an apple from the wagon for him to sniff. He got him going once more down the little hill to get speed for starting up the long incline ahead. He kept the apple close to Hank's nose and maintained a brisk pace. As they reached the steep part of the trail again, Hank began to slow somewhat and Clark let him get a small bite of the apple. After a few more steps, he gave him another bite.

This continued until they were near the top of the hill when Hank took the last bite of the apple. Fearing that he would stop again, Clark looked around for a good strong stick. Spotting one beside the trail, he picked it up and held it where Hank could see it. "You better not put me through this again, Hank. If you stop, you are going to get a terrible lick on your head."

He picked up his pace somewhat, hoping Hank would recognize how close they were to the hilltop. He didn't know what worked but gave a long sigh of relief when they finally reached the top. Starting down the other side, he stroked Hank's neck and said, "Good job, old fellow."

Clark was half an hour behind the rest of the caravan reaching Chastain's Station. Jimmy rushed over to him. "Boy, did you have me worried! I wondered if we would have to leave you back there through the night and take another mule back in the morning to pull the wagon up that hill."

"Well, it's a good thing Sam Pickens gave me the hint about what to do if Hank balked," Clark said. "Even taking his advice, I almost didn't get him to pull the wagon to the top of the hill."

Sergeant Adderholt called to them, "Fellers, we have a little cot in the barracks you can sleep on tonight. It'll beat sleeping on the ground." They gratefully accepted his offer.

They bedded down soon after eating and fell asleep quickly, but shortly before midnight they were awakened by screams from outside. Leaping up, they rushed to see where the sound was coming from. It was the tent where Muata's family was staying.

Muata rushed toward them. "Get help! My sister's baby is being born!"

The Army medic came hurrying with a lantern and medical supplies. He called for someone to get a fire going to heat water. "I need towels and sheets, too," he said.

The whole camp waited anxiously as the woman's screams pierced the night. Finally, the screams stopped and they heard the tiny cry of a newborn baby. The medic came out and announced, "It's a girl."

Muata rushed to his side. "Will my sister be okay?"

The medic nodded. "It's better the little one arrived here at the stockade rather than later when we are on the trail traveling west. She will be able to gain some strength in the weeks we're here."

Clark and Jimmy went back to bed but found that sleep wouldn't come readily after the crisis had raised their adrenaline. They were still groggy when Sergeant Adderholt shook them awake the next morning. "Come on, boys! We've got to get the wagons unloaded."

They spent Wednesday placing the contents of the wagons where Adderholt wished, and Thursday they made lists of the Indians who occupied each of the tents in the stockade. Because Muata spoke English, Clark asked him to help them with making the lists. He wondered how they could have managed to do it without him.

Friday dawned with a steady downpour of rain. Clark and Jimmy hitched the mules to the wagons and headed for home. Muata watched dejectedly as they left. They waved goodbye to him, feeling deep sorrow at the sad plight of him and his family.

Part 2

1839 – 1863

Happiness lies in the joy of achievement
and the thrill of creative effort.

Franklin D. Roosevelt

13. Kentucky Trip

Clark eagerly slit the envelope open when he saw the return address: Edgar Carmichael, Louisville, Kentucky. He was oblivious to the inquisitive eyes of the postman as he removed the thin sheet inside and started reading the reply to a letter he had written asking Edgar when it would be convenient for him to pay a visit and see the gliders he had built.

Dear Clark,

I would be most happy to have you visit whenever it is convenient for you. Sampson Picklesimer told me how very much you were interested in flying when he met you at your uncle's house last winter -- the time the snowstorm halted his trip. Plan to stay a few days up here. I can show you how I build a glider, and you might even take a flight on one of them. I will be looking to see you before the weather turns cold again.

Regards,
Edgar Carmichael

Clark folded the letter, put it in his pocket and turned to leave the store.

"Wait a minute. I have another letter addressed to your Pa that you might want to take with you," the postman said. He handed Clark the letter and waited to see if he would volunteer any information about the letter from Kentucky. Getting no further response from Clark, he threw the mailbag over the horse's shoulder, climbed astride its back, flapped the reins, and started up the trail toward his next mail stop.

Clark headed home, wondering if he should plan to take the trip to Kentucky alone or ask Jimmy to go with him. He didn't like the idea of going by himself, but he feared Jimmy would be a distraction, for he wasn't curious about how things are built, nor could he be satisfied with spending hours perfecting some complicated mechanical part. And Clark knew his brother surely wouldn't be interested in taking a flight on one of the gliders.

As he walked along the trail that followed the curve of the valley, his head was filled with plans and dreams. Yet, even with the mental distraction, he took notice of the rows of tasseled corn-stalks in the field, bearing ears with silken tips that would soon be ready to harvest. He listened to the harmonious sound of the creek as it splashed over the rocks, and smelled the fragrance of the wild asters, ironweed, woodbine and snakeroot growing on the banks. He sensed the moving and maturing of all nature around him on this mid-September day, making him wonder how anyone could find life boring or meaningless with so much to observe and discover around them.

When he reached home, he found Pa sitting on the edge of the porch with his head leaning against a post. He looked pale and his eyelids were drooped. Clark ran to his side.

"Pa! Pa, are you okay?"

"Yeah," Elisha murmured as he pulled himself upright. "I guess I nearly blacked out. I was cutting hay down in the meadow all morning and I must've got too hot."

"Let me get you some water," Clark said, rushing into the house. He dipped a cup into the water pail on the little washstand and hurried back outside, placing it carefully in his Pa's hand.

"I'll help you over to the shade," he said.

He settled him in the shade of the big oak beside the porch and fanned him with his straw hat, thankful to see some color returning to his face as he sipped the water.

"Where's Ma?" Clark asked.

"She took Lump and the girls and went looking for grapes along the creek bank."

Clark reached in his pocket and brought out the letters the postman had delivered.

"You got a letter from Uncle Joe today, and I got one from Edgar Carmichael. I've already read mine. Mr. Carmichael has invited me to come up and see his gliders. I would like to leave next week if it won't cause a problem for you. I'll have to be gone about two months to travel up there, spend a few days with him and return. Will that be all right with you?"

"Yeah, that'll be fine, son. It's probably the best time of the year for you to do it, and I know how you've been looking forward to going." Opening his letter, he said, "Let's see what's on Joe's mind."

He finished reading the letter, folded it and tucked it in his shirt pocket. "He's interested in knowing how our crops fared this year and whether we're going to bring a load of goods down to Gainesville this fall. He'll be happy to hear what a good year we've had with the crops. A lot of folks in town, who don't live off the land like we do, have been going through hard times since the Panic. We probably won't be able to get good prices for our produce like we did last year."

"Are you feeling okay now, Pa? Why don't you let me finish cutting the hay? You don't need to be working in this hot sun any more today."

When Elisha willingly handed over the sickle to him, it caused Clark to study his face for a moment as he tried to determine whether he truly was feeling all right. It was so unlike him to concede that he should stop a job in the middle of the day. But he looked normal to Clark, so he tossed the sickle on his shoulder and headed for the meadow.

After an hour of cutting and shocking hay, Clark had perspiration running down his face. No wonder Pa nearly passed out, he thought. There was no breeze across the field and the high humidity was intensifying the heat. He walked over to the little creek and bent down to rinse off in the cool water. As he rose and wiped his face with his shirtsleeve, he saw movement in the bushes on the other side of the stream. Looking closer, he saw that it was a boy.

"Hello," he called.

The boy shrank deeper into the bushes and Clark waded across the creek toward him. As he got near, the boy sprang out of the bushes and fell at his feet. He was a very thin Cherokee lad about seven or eight years old.

Clark reached down and patted his shoulder. "Do you speak English?" he asked.

The boy looked at him with fear and pleading in his dark eyes. Clark knew some of the Indians had hidden in the mountains to escape capture when the Army came to round them up and take them away, but this was the first one he had seen in the three months since the removal.

The boy said something to him in Cherokee. Clark replied, "Young fellow, we'll have to depend on sign language to talk to each other because I don't understand what you're saying."

The boy made signs of eating and pointed up toward Cedar Mountain. Clark interpreted this to mean that he had relatives hiding on the mountain who were in need of food. He motioned to him to follow and started toward the house to get food to send with him.

Arriving at the house with the Cherokee boy in tow, Clark took him to the back door. It will be better if Lump and the girls didn't know about this, he thought. If they were to tell someone about it and word reached the government that we are helping Indians hide, they would put all of us in jail. He put up his hands toward the boy to show him that he was to stay there at the steps.

Going inside, he found Ma in the kitchen. He quietly told her about the boy and asked if she had something he could take for the boy's family to eat. She had bread that she put in a bag and a pot of pinto beans with pork ribs that she put in a tin pail for him to take. He went outside and handed it to the boy who smiled and bowed to him, then quickly ran into the woods.

Clark said, "Ma, I guess you will be having to find food for them every few days now that they've located a willing helper."

"I'll be happy to give them food as long as we have anything we can share with them," she said.

"Ma, did Pa tell you that I am going to Kentucky to see Edgar Carmichael, the man who builds gliders?"

"No, he didn't mention it. When are you going?"

"I plan to leave in the next few days. I got a letter from him today saying that now is a good time for me to come."

"Bluford and Cager have been planning to go up there sometime soon to see the part of our family that stayed behind in Kentucky after Grandpa Bluford died in 1816. In the twenty-five years that have passed since his death, we haven't seen any of them. We've kept in touch with some letters, and that's all. I hope the three of you can go up there together. I don't like the idea of you traveling that far all alone."

"Okay, I will talk to them, and I hope it will work out that we can go together. I didn't want to take Jimmy because he really needs to stay here and help you and Pa while I'm gone. He wouldn't like going there anyway. It's a long time to be gone and he would hate spending several days working on gliders with me and Edgar."

A few days later, Clark had his bag carefully packed and was ready to leave on the trip with Bluford and Cager. He had stashed a tablet and pencil for writing and sketching everything that he would learn while he was in Kentucky. He wondered what kind of materials Edgar used for building his gliders and how he launched them. It was going to be exciting to spend several days talking and working with someone who had the same keen interest in flying that he had.

"Boys, I've packed this basket of food for you to take with you; it's a long trip," Ma said as she came from the kitchen and handed the basket to Bluford. "I know you all will be careful, but I won't rest well 'til you're back home. Tell our kinfolks I'd like to see them and would be happy to have them come down and visit us whenever they can."

"Ma, you keep an eye on Pa and don't let him overwork like he did last week," Clark said. "He tries to do too much for a man his age."

"Yes, he does. If I catch him overdoing, I'll threaten to take him to see Doc Anderson. That ought to stop him," she said with a chuckle.

The young men each gave her a hug and climbed aboard the wagon. Ma's a little woman, Clark thought, but she's a force to be dealt with when she makes up her mind about something.

Twelve long days on the trail had brought the three travelers to a few miles north of the Tennessee-Kentucky line. It would take them another ten days to reach Louisville on the northern boundary of the state. Although they were weary, they felt that the second half of the trip would be somewhat easier, as they could take the trails traversing the Pennyroyal Plateau. The farmers and traders who used those trails had kept them smooth and well-trimmed. The people in the area were friendly and would likely welcome them to spend the nights in their barns. Some of them would perhaps allow them to get apples, pears, walnuts, and the like, from their trees as they passed through.

"I tell you what would be nice," Bluford said, "that is to come across a good cook that would make us a big pan of cornbread. I can just taste it now."

"I'd like a bowl of chicken and dumplings," Cager said.

"As long as we're dreaming," Clark said, "I'll take a blackberry cobbler. Meanwhile, I'll look around for some fallen limbs to make a fire so we can cook some taters to eat with our deer jerky."

Bluford reached in the wagon for the frying pan and a bucket of lard they had brought. Cager started a fire with small twigs that were around the trees where they were camping for the night. Clark returned shortly with an armload of limbs, which he piled on the little fire, and soon they had their simple little meal ready to eat.

A full Harvest Moon was rising over the mountain as they settled down for the night. A whippoorwill was calling in a nearby pine, and farther down the valley another was answering. Clark pulled a blanket over his shoulders and settled his head in the crook of his arm, glad they had found a good camping spot for the night.

Their calculations of travel time turned out to be quite accurate, and they arrived in Louisville on October 6th at the home of their Dyer relatives.

"I believe I would have recognized you young men even if I hadn't got the letter from your mama saying that you were coming to visit," declared their Uncle Hadley as he came out to greet them. "All three of you bear a resemblance to your daddy, though the youngest there not quite as much as you older ones. Twenty-five years! That's how long since I have seen your daddy and his family. How is he doing?"

"He's doing pretty well," Bluford said, "except for a little trouble with his heart from time to time."

"Well, come on in and let's see if Mary Ruth has supper ready for us."

Their aunt came from the kitchen wiping her hands on the tail of her apron. She was tall and thin, and she greeted them with a broad smile.

"I bet I can guess that after three weeks of traveling you boys are ready for a big home-cooked meal," she said. "There's a wash stand on the back porch where you can rinse off before you eat."

As they went through the kitchen to the porch, they sniffed the air brimming with the aroma of food she had cooked for them. They swapped glances and happy smiles in anticipation of the good meal they were about to have.

When they were seated around the table, Hadley said, "Let's say grace." All heads bowed immediately.

"Lord, we humbly accept this bountiful meal that thou hast given us this evening. We thank thee for providing it, and thank thee for safe travel for our nephews over the past weeks. We ask thy blessings upon all of us and for thy leading and direction in all that we do and say. We ask these things in the blessed name of Jesus. Amen."

"Amen," they all echoed.

Bowls of food were passed around the table, and the young men dipped generous portions onto their plates. They ate heartily, and Aunt Mary Ruth looked on in satisfaction.

"It's a pleasure to have you boys come and visit," she said. "We enjoy the letters we get from the family, but it's so good to see some of you face to face again."

"Aunt Mary Ruth, I can't tell you how wonderful this supper is," Clark said. "I'm afraid we've made pigs of ourselves."

"Don't you worry about that for a minute," she laughed. "I would have been disappointed if you hadn't eaten heartily. I cooked extra food in the hopes that you all could make up for the scanty eating you've had as you traveled."

"We've arranged for the rest of the relatives who live around here to meet us at the boat dock in the morning," Uncle Hadley said. "We thought you all would like to see some of the big steamboats that travel down the Ohio River taking tons of crops and livestock products to New Orleans and other markets along the Ohio and Mississippi rivers. We're planning to do a little fishing in one of the lakes, too."

"I'm really looking forward to seeing the river," Bluford said. "Even the Indians that lived in the Georgia mountains near us talked about the Great River."

"What I'm wanting to see is a big steamboat," Clark said. "If we have time, I'd like to go onboard one."

"If they have one docked that's not being loaded or unloaded, I will ask about us going aboard," Hadley said.

"I think some of us told you in our letters that I am planning to spend a few days with a man up here, Edgar Carmichael, who builds gliders. I'm planning to go to his house day after tomorrow."

"Yes, we knew you were planning to spend time with Edgar while you were here. We know about his gliders. The newspaper has stories about him from time to time. Don't know that I'd be interested in taking a flight on a glider. How did you happen to get interested in flying, Clark?"

"I've been interested in how things work about all of my life. When I see the big birds flying over the fields, gliding on the wind currents, and coming in for a smooth landing, I just believe that somehow a man can build something that he can fly in a similar way."

"You may be right. I know Edgar thinks so. You just be sure you know how to control one of those things if you decide to fly on it."

"I'll do that, Uncle Hadley. I don't want to break my neck…that would ruin my dream of building a flying machine." They both chuckled at that.

"Well, fellows, let's hit the sack so we can get started early in the morning," Hadley said.

The travel-weary nephews didn't need a second invitation to head for their first night of sleep in a real bed in nearly a month.

14. Ohio River Sights

The three young men rose early the next morning, but their hosts had already been up about an hour ahead of them. Uncle Hadley had fed the livestock and milked the cows, and Aunt Mary Ruth had cooked a full breakfast for the family.

Right after eating, all five of them climbed into the buggy and their spirited horses seemed to be as eager as they were to get on the road for some adventure. They traveled directly to the boat dock and Clark and his brothers were agog at the sight of a river so wide. Dozens of boats were docked there, and a multitude of people were working and milling around.

Hadley called out to a man standing on one of the large steamboats at the dock, "George, I have some visitors who would like to take a look at your boat if they won't be in your way."

"Bring them onboard," he called back.

They quickly headed for the gangplank and boarded his boat. Two burly men were feeding a large furnace that provided steam for the turbines powering the stern-mounted paddle wheel. Smoke billowed up from a tall pipe that was exhausting smoke from the furnace. It was a plain boat, used only for transporting livestock and produce, but up the river they could see a fancy steamboat that was taking on well-dressed passengers.

A loud horn sounded and soon they saw one of the steamboats begin to pull away from the dock. They watched in amazement as the paddle wheels churned the water and the captain steered the moving boat into the river channel.

"That boat will travel a thousand miles in a month carrying a load downstream," George told them. "Even making the return trip upstream it will cover the same distance in three to four months. That's a great deal faster than a loaded wagon can travel on land. And the boat can carry a much larger load than any wagon can. With increased speed and larger loads, it would cost a customer three times as much to transport goods by way of wagon as by boat if they're shipping upstream and twenty-five times as much when shipping downstream."

"Wow!" Cager said. "No wonder there's so much activity here at the river. Why wouldn't any sensible businessman ship by steamboat with that kind of savings?"

Clark was impressed by the steamboats in a completely different way. "I'm amazed," he said, "at how clever the man was who created a way for the steam turbines to do the work of rowing the boat by powering the paddle wheels to do it."

"They're using that same method to power railway locomotives in Maryland and Virginia," George said. "I foresee it becoming the chief method of transporting people and goods across long distances in just a few more years."

Hadley walked across the deck and leaned on the hull of the boat, watching his nephews with amusement as they took in this previously unknown world of commercial activity.

Finally, he said, "As soon as you fellows have seen all you want of the boats, we'll head over to the lake to meet your other relatives and do some fishing. I also want you to see the Falls of the Ohio. It's not like Niagara Falls, but it's pretty impressive when you see the waters of this mile-wide river fall about ten feet at one point."

They were reluctant to leave such an interesting place but knew they needed to get going in order to fit in the other activities their uncle had planned for the day. As they walked to the buggy, Hadley proudly told them about the tons of produce from Kentucky that was shipped down the river—tobacco, livestock, hemp, corn, rye and other grains.

"But the most important crop we have is hemp," he said. "We're producing nearly all of the nation's hemp which, you may know, is mostly used to make rope."

"Actually, we grow all those same things on our Georgia mountain farms," Bluford told him, "but the acreages of your farms here in Kentucky are so much larger than ours. It is so mountainous where we live our valley farmlands are quite narrow compared to yours."

"Folks who've been down there to visit tell me that it is really pretty country. I'd come down to see you all if I wasn't so tied up here taking care of the plantation."

"I see that you really have a lot to do, but if you can find somebody you trust to take care of it for a couple of months during the slow times, maybe you can take the trip to the Georgia mountains to see us, Uncle Hadley."

When they got back to the buggy, Clark was leaned against it with his pencil and tablet in hand. Bluford looked over his shoulder to see what he was writing and saw that he had sketched the stern-mounted paddle wheel of the steamboat that they had seen at the dock.

"I guess you're making plans to build one of those things when you get back home," Bluford said a little sarcastically.

"Well, I don't think one would be of any use on our little creeks," Clark said, "but there's no reason why a man couldn't put that type of design to good use for other things."

He thought to himself, I'm not sharing any ideas I have with him about ways I might tailor the method to control the airflow across the wings of a flying machine. It would be a waste of breath to try to make him understand and would only bring on more of his ridicule.

Aunt Mary Ruth was waiting in the buggy for them. She had been very patient, seeming to understand that all of the sights and discussions were making memories that would last a lifetime for the young men. She smiled at them as they climbed into the back seat.

"You all should have planned to spend at least a month so you could see more of this country."

"Yes, Ma'am, that would have been nice but cold weather may overtake us even during this short visit," Bluford said.

Hadley had been talking with a man whose buggy was parked beside theirs just before he climbed in to start their trip. "You mentioned the cold, Bluford," he said. "The man I was just talking with over there said we might get an early snow here next week."

"Oh, no!" Mary Ruth cried. "We have lots of late crops to harvest before winter weather comes."

"We'll have to get busy and round up all the helpers we can find to gather in the produce. It could be a false alarm, but I'm not willing to take a chance on losing our crops."

They started along the well-packed road and traveled in silence for a while, each with his own ideas and dreams bouncing around in his head. Eventually, the rhythmic thumping of the horses' hoofs brought their wandering thoughts back to the present moment and place.

Looking back at Clark, Mary Ruth said, "I notice that you have a strong interest and aptitude in all things mechanical. Do you plan to work in something where you can use those skills?"

"I would like to," he replied, "but I don't know if there will be anything available for me in our area."

"What would you think about moving to a larger town where there are several different industries hiring men with good mechanical aptitude?"

"I enjoy visiting places where a lot is going on, Aunt Mary Ruth, and I pick up good ideas from them, but I really wouldn't want to live in such a place. The little hidden mountain valley where we live is just right for me. I am the most creative when I am there experimenting with any new notion I get about how something would work. I don't know how to explain it, but my mind seems clearest and most imaginative when I am apart from hustle and bustle, just alone with the wind blowing through the trees and the creek splashing over the rocks."

Mary Ruth laughed appreciatively. "From your description of the place, I also believe that anyone could think more creatively there." Turning serious, she added, "I hope you will have the opportunity to do something special with your talents, Clark. God wants every one of us to use the talents he gives us and to multiply them."

"We're coming up on the Falls of the Ohio," Uncle Hadley announced to the group.

They could hear the roar of the river as it fell over the rock outcrops for some time before it came into their view. Finally, the buggy turned a curve in the road and passed through some thick bushes. There it was, the vast span of the water and the spillway running almost the width of the river churning and spraying up a mist.

"Holy smoke," cried Cager. "That is something to see!"

They all gazed at the river, mesmerized by the sight of the sparkling water and the crashing waterfall.

"There are some smaller falls a little ways down the river, but this is by far the biggest and most spectacular one," said Hadley. "This is the only portion of the river that is too rough to be navigated. The river runs nearly a thousand miles before meeting the Mississippi River at Paducah. As you might know, from Paducah the Mississippi flows all the way down to New Orleans, then from there to the Gulf of Mexico. I would love to take a ride on a steamer the whole way."

"Me, too, Uncle Hadley," Cager said. "I've heard some of the travelers passing through Choestoe talk about spending time in New Orleans and boating on the Mississippi River, or 'the Great River' as the Indians call it."

"Maybe it will be possible for us to make the trip together sometime, Cager," Hadley said as he turned the horses back onto the main road and headed toward the lake.

When they arrived at the lake, some of the family members were already there. Some were fishing from the banks and a few more were on the lake fishing from a boat. The women and children were starting a fire and unloading pans and cooking supplies.

"Boys, you all get yourselves fishing poles off the hooks on the side of the buggy. There's bait in the bucket back there. Mary Ruth brought a frying pan and lard. She's going to be ready to cook whatever you catch," Uncle Hadley said.

They wasted no time getting poles and baiting their hooks. Uncle Hadley walked with them over to the lake and began introducing them to the relatives, who were very welcoming and interested to hear about the families back in Georgia. Happenings around Louisville had also generated a lot of talk.

"A boatload of slaves escaped last night. They crossed the river to Indiana," Charles, one of the relatives, told Hadley. "It seems that somebody from the Underground Railroad has been passing information to the slaves on the plantations around here that they will help them escape to free country if they will get a group together and let them know when they are ready to go. I hear that every one of David Swift's field workers escaped with the group last night."

"David bought two new slaves at the auction in the spring," Hadley said. "It's likely they're the ones who instigated the escape plan. The rest of the slaves have belonged to David's family since he was just a boy. It will take a lot of time and money to build a workforce again that's large enough to manage his plantation."

"I suppose those slaves who've been with the Swift family so many years are going to be disappointed at what life is like for them when they have to make it on their own," Charles said.

"I fear the controversy over slave ownership is going to bring this country to war. Plantation owners need the certainty of a workforce, and the majority of the slaves are glad to have the assurance of food and shelter being provided for their families in exchange for their services. The Northerners don't understand that most times this is a good arrangement for both parties."

Clark, Bluford and Cager were surprised to hear this discussion. Their family didn't have slaves and neither did their Choestoe neighbors. Work on their farms was done by the parents and children, with extra help being hired when it was needed. The idea of people being bought and sold to do the farm work and not having the freedom to come and go as they pleased seemed loathsome to the boys.

It didn't take long for the fishermen to catch enough fish to bring to the picnic site for the women to begin cooking. Cloths had been spread over the rough tables and anchored with plates and cups to keep the wind from blowing them off.

The children had drawn a pattern in the dirt and were playing hopscotch. Clark thought of home and how much Lump, Melinda and Matilda would enjoy being here at the lake playing with the other children. Jimmy and Lige would have loved the steamboat visit and the fishing and picnicking here at the lake. He wondered if Pa had had any further heart problems and sent up a quick prayer for him and Ma.

As the smell of food began to drift across the area, everyone started moving toward the tables. When the last of the fish was cooked and served, Uncle Hadley asked a blessing and they began eating.

"Do you fellows do much fishing down your way?" someone asked the Georgia boys.

"We do some trout fishing but we don't have many cookouts like this," Bluford said. "This is certainly good eating."

"Did I hear that one of you is going to see Edgar Carmichael and his gliders?"

"Yes, I'm planning to go over and spend tomorrow and the next day with him," Clark said.

"How did you get interested in gliders?"

"A traveler passing through Choestoe told me about Mr. Carmichael and I decided I would like to see his gliders. I love building things and I want to build a flying machine myself."

"Really? You're going to build one by yourself?"

"I probably will. No one else seems to be interested in it."

"Actually, he takes a lot of ribbing about his flying machine idea," Bluford said. "Folks think it's very odd for him to spend a lot of time working on something like that."

"Well, I wouldn't worry about that, son. If you've got the talent and interest, I say do it!"

Clark smiled slightly and nodded, glad to get encouragement from someone. Tomorrow should give him a better idea about whether his dream of building a flying machine was realistic or not. He hoped the weather would be good for gliding with Mr. Carmichael.

15. Kentucky Gliders

Clark was excited as he headed out the next morning to see Edgar Carmichael who lived about ten miles southwest of Uncle Hadley. His uncle had let him drive his buggy for the trip, and it was a welcome change from the wagon he and his brothers had ridden from Choestoe to Lexington. Hadley's buggy had a padded seat with springs to absorb most of the bumps. The horse was young and spirited, and the scenery was beautiful as he traveled along the road toward the mountains. The miles passed quickly and pleasantly, bringing him to Edgar's house in a little less than two hours.

He brought the horse to a halt at the front gate. A pair of hounds roused on the front porch and came toward him barking and wagging their tails. In a few minutes the front door opened and a husky, curly haired man came out, ordering the dogs to be quiet and sit down. He strode to the buggy with an outstretched hand.

"Edgar Carmichael here," he said, as he gave Clark a firm handshake.

"I'm Clark Dyer, Mr. Carmichael. Glad to meet you."

"Oh, call me Edgar. You'll make me feel old if you call me Mister," he laughed.

Taking the horse's bridle, he said, "I'll lead you around to the barn where we can unhitch the buggy and put the horse in the pasture. My workshop is back there, too. I'll show you around. How was your ride over here?" He was looking the buggy over appreciatively.

"It was very pleasant. This is my uncle's buggy and a great improvement over the buckboard my brothers and I had for our trip up from Georgia."

"I'm sure that's true. I bet the road coming here was lots better than some of the mountain roads you had to travel, too."

"Yes, indeed, it was. With a fine buggy and the beautiful scenery in this area, the ride over here was fast and pleasant."

As they walked into Edgar's barn, Clark's eyes immediately fell on a glider parked near the front. The wings on each side looked to be about seven and one-half feet long, curving upward in the center like a bird in flight. They were joined together with a pole about fifteen feet long and were anchored by a cross pole. A harness was attached with leather straps to each of the wings and dropped down to form a pad for the pilot to recline upon while in flight. There were also straps for the pilot's arms that were fastened to the wings and poles that went downward to fasten to wheels at the front.

"You're going to think you've become a bird when you get airborne with this glider," Edgar said, smiling broadly.

"I hope you will give me some guidance first," Clark said. "I would hate to disgrace you by breaking my neck on my maiden flight."

"There's not really a lot to learn. I'll ride with you first. Then later, when I explain how the thing is built and how it works, it will make a lot more sense to you."

"This looks new. Did you build it yourself?"

"I did most of the work, but I had to call in some help for setting the wings in place. I helped another fellow build one who lives not too far from your Uncle Hadley, so he was happy to come help when I needed him."

As he talked, Edgar pulled the wings downward on each side of the glider to make it easier for them to pull it to the launch site.

He brought the harnesses on the glider forward and handed the left one to Clark, keeping the right one for himself.

"Let's tow this bird out to the hill and see if it's a good day for flying."

A few hard tugs by the two of them started the glider rolling out of the barn, and they pulled it along a path leading up a lengthy ridge. The wind was quite brisk and it kept whipping against the wings, compelling them to hold tightly to the harnesses to keep the glider in the path.

"We should be able to soar today with this much wind," Edgar said. "I wish it weren't so blustery, but maybe it will smooth out by the time we're ready to lift off."

Clark looked at the glider and wondered if it would be sturdy enough to hold up both of them. It appeared that the cured willow framework had been securely attached to the canvas stretched over it. He could see the top of the hill ahead of them and estimated its height to be about eighty feet. It would be a nasty fall if they crashed, but with good luck perhaps they could land in position to roll down the hillside, he thought.

"After we're airborne, lean with me when you feel my weight shift to the left or right," Edgar said. "That's how we will direct our turns as well as our ascent and descent. I will keep the nose up when we're ascending and level out when we reach a good altitude. When we get ready to descend, I will bank into the wind to produce drag to slow us for landing. If we happen to hit an updraft and get a sudden lift, pay close attention to my movements and follow suit. It will probably be a short thrust but it will definitely scare you the first time it happens, maybe the second and third time, too," he laughed. "I will try to judge the wind carefully so we won't get any bad surprises."

"Do you have any questions?" he asked Clark as they neared the top of the hill.

"Not right now. Your explanation makes sense to me. I will try to follow your instructions. It's really exciting to be about ready to take off."

"We'll rest here a little while from our haul up the hill before we embark. Keep a firm hold on the harness since the wind is coming across the hill pretty strongly. We don't want this bird to take off without us, do we?"

Edgar turned the curve of the wings away from the wind and sat down against one of the wheels to brace it. Clark stood holding the harness on the other side, bracing himself against the pole fastened to the wheel. He looked across the valley below and tried to imagine what this adventure was going to be like. He glanced over at Edgar and saw that he wasn't experiencing any anxiety whatsoever. This is what I came here to do, he thought, and I think Edgar is the right man to fly with. Ma would die if she could see me here.

God, keep us safe, he prayed silently.

Shortly, Edgar stood up and lifted the wings in position for flight. He slid his arms through the harness and said to Clark, "Fasten the harness around your shoulders like this," as he showed him how. "Now, we'll bring this strap across both our chests to hold us firmly to the frame, and we will start running across the hill into the wind. As soon as you feel your feet come off the ground and we begin rising, lift your legs backward and rest your feet on top of this bar," he said. "I will do the same. As I told you earlier, lean in the same direction I do while we're airborne and, if all goes well, we will keep the same position until we finish our flight and descend to the ground. We should roll to a stop before we put our feet down again."

He looked over at Clark. "Are you ready to roll?"

"I'm ready," Clark replied.

They began running and Clark was surprised at how quickly his feet began lifting from the ground. First it was a short bounce, then the glider started to rise rapidly. Clark's breath was caught away when he found he could no longer touch the earth. The forward movement swept their bodies backward and Clark lifted his legs higher in search of the footrest. Finding it, he hooked his feet in place.

Looking at the valley spread below them, they floated quietly above the earth with only slight bouncing and rolling as the updraft carried them along. The movement felt as gentle as the rocking of a baby's cradle. Clark thought: so this is how it feels to be a bird. He wished the flight could go on all day. The effects far exceeded anything he had ever imagined flight would be like.

He felt Edgar lean to the left and he did likewise. The glider banked and he felt it rise higher as the wind hit the wings at a different angle. They tilted their weight back to the right and leveled out again. Clark estimated they were about twenty feet higher now and far above the altitude he'd expected them to reach. His excitement was soaring even higher than the glider.

Suddenly, a strong updraft hit the glider and gave them a rough boost. He felt Edgar tense as he leaned left again. This time their turn was not as smooth, and the glider started rocking back and forth, then began to quickly lose altitude. Edgar leaned steadily left and Clark did the same. They held their position firmly for several minutes, and finally the shaking stopped as the wind hit them more evenly. Edgar eased the glider back to level again.

"I think we'll go down now," Edgar said. "The wind is too unpredictable to continue any longer."

He began banking right and slowly pulled the front downward. The maneuvers brought them out of the center of the wind's channel and the glider began to descend slowly.

When they were once again floating smoothly, Clark felt his heart finally stop pounding. He could plainly see that there was more to flying than just knowing the principles of lift and drag, which a pilot could produce by turning the glider in different directions to the wind. There were also variations in updrafts to be dealt with that could not always be anticipated, and these could send an aircraft plunging to the earth unless the pilot speedily regained control.

As they floated toward the earth, Edgar steered the glider to a smooth strip in the pasture near his barn where the wheels soon hit the ground with a thump. Clark found his legs to be a little wobbly as he lifted his feet off the bar and stood up.

"Well, what do you think?" Edgar asked. "Is this something you're still interested in doing after the shaking we had up there?"

"Oh, yeah. It was really scary for a few minutes until you got the glider back under control, but it was a great thrill to be sailing through the air like a bird. I know I've got a lot to learn before I can build and fly one of these things myself, but I will do it. The flight confirmed to me that many of the thoughts I've had about how I should build one are right on target."

"Help me fold the wings down and we'll pull this thing back to the barn. I'm starting to build another one and I'll show you how I'm putting it together. Every one I make is a little different from the others. I try to make each new one with a better design than the previous one."

"Where are the others you've built?"

"I sold the second one I made but used most of the parts from the third one to make the fourth. I will always keep the first one I made. It's stored in the shed over there. I'll take you to see it before we go to the house."

They pulled the glider into the barn, parked it, and then headed to Edgar's workshop in back. He had already constructed some of the frame for the new glider. He had canvas cut and ready to be sewn across the frame for the wings. With this design, there wouldn't be two separate wings. Rather, the canvas would stretch all the way across to cover both sides of the frame.

"I will make the canvas airtight with applications of pitch," Edgar said.

"Did you have very many crashes when you first began flying?" Clark asked.

"Yes, I had quite a few and I actually had to rebuild my first glider a time or two it was so badly damaged in my wrecks."

"Were you ever hurt badly in a crash?"

"Yeah, I've banged myself up a couple of times," he said with a laugh. "Broke a leg once and an arm another time. I've had some pretty bad bruises, too. But you learn how to position your body so the frame takes most of the impact when you realize that you're going to crash. But I've been lucky, too. I know that some of my close calls could easily have been fatal. I guess it just wasn't my time to die."

Edgar smiled wryly. "I often invite the newspaper reporter to come out and see my flights, and he seems to especially enjoy writing the story when I have a crash. Actually, I sometimes think everyone else relishes the stories better, too, when they're telling about my failures instead of my successes."

"I know what you mean about that," Clark said. "My brothers and their friends take such delight in poking fun at me about my plans to build a flying machine. When something goes wrong for me, they just whoop and holler in their pleasure over my failure. It means a lot to me to get to meet you and talk to you, Edgar, since you understand what it's like to be in the same situation I am in."

"Well, don't let their reaction get to you, Clark. Always remember that your work is just as good when someone ridicules it as when they praise it. The important thing is for you to be satisfied in your own mind that you've done a good job. Most times there are things any of us can see that we could have done better after we get a project completed, but we just have to aim for improving those things the next time around."

He slapped Clark's shoulder. "Come on," he said. "Let's head over to the shed. You need to see the first glider that I was able to get airborne. It looks very amateurish, I know, but I am very proud of it."

Clark was surprised at what he saw when they reached the shed and Edgar opened the door. The little glider was barely more than half the size of the one they had flown that morning. There were two separate wing frames joined together with a cured willow pole for its spine, which extended at the front about two feet past the wings. It had a crossbar that stabilized the wings. The only other part it had was a harness to fit across the pilot's chest.

"How far were you able to fly with this?" Clark asked.

"When the wind was just right, I sometimes floated at an altitude of about twenty-five feet for a distance of around two hundred feet. It doesn't sound like much, but I was jubilant at being able to do that. Since it was my first time to be airborne, I was thrilled to death."

"I can understand that. I guess since you were only about twenty-five feet high you could drift to a pretty smooth landing, couldn't you?"

"Yes, especially when the wind was blowing just right." After a moment of silence, he said, "Let's head over to the house and see if the missus has supper ready."

As they reached the porch, they didn't have to wonder any longer if Mrs. Carmichael had the meal ready. The smell of fried chicken and biscuits was drifting out the open kitchen window.

"Cora, come meet our visitor from Georgia," Edgar called as they entered the front door. She came bustling into the room, smoothing her apron with one hand and her light brown hair with the other. Her cheeks were flushed from cooking over the hot wood stove.

"Meet Clark Dyer, Honey," Edgar said. "Clark, this is my wife, Cora, the little lady who has put up with me for over twenty years."

"Glad to meet you, Mrs. Cora," Clark said, as he shook her hand.

"Happy to meet you, too, Clark," she said. "I have already heard that you are as fascinated with flying as Edgar is."

"Yes, ma'am. It's so kind of you all to invite me up to have a first-hand look at the gliders and especially to get to take a flight on one. I am very grateful to you."

"We're happy to have you," she said. "Come on in and make yourself at home. I will have supper ready in a little while."

"Clark, you can walk down to the creek if you'd like to while we're waiting for Cora to finish cooking. The kids are down there somewhere, probably fishing. There's some mighty fine trout in that stream but they're a bit skittish," Edgar said. "I'm going to feed the cows and horses in the meantime. Tell the kids to come on up to the house for supper."

Clark ambled down the path leading to the creek, still thinking it felt like a dream to actually be here, seeing the gliders and finally getting to fly on one. As he neared the creek he saw a young girl sitting on the bridge over the creek. She appeared to be about his age. She seemed to be in deep thought and didn't notice that he was approaching. The setting sun against her back formed a golden halo around her auburn hair as she poked at some dragonflies floating in the water. He stood for a few minutes hating to interrupt her daydreams, but finally felt he needed to make his presence known.

"Hello," he said. "I was told to come down and tell you to come to supper."

His voice startled her and she swung around, throwing her stick into the creek.

"I'm sorry to spook you. I'm Clark Dyer from Georgia and I came up to see your dad about his gliders. Can I help you get up?"

She scrambled to her feet and he saw that she was about five and a half feet tall with an athletic build. She was barefoot and wearing a floral dress that reached to her ankles. She visibly calmed as she saw that he was a pleasant-looking young man who was perhaps no older than she.

"Hello. I'm Betsy. I heard Daddy telling Mama that you were coming, but I expected you to be a lot older than you are," she said.

"Let me guess. You thought I would be older because you thought only older men would be interested in building gliders."

"Yes, that's right. There are not many men of any age who are interested in doing that. Most people around here wonder why Daddy wants to do it."

"Well, he's a very talented man and I am happy to get to spend a couple of days with him and learn as much as I can about how he does his work. Have you ever flown on one of his gliders?"

"Absolutely not! Neither has anyone else in the family."

Clark was reminded again of how her family's disdain toward Edgar's flying was so much like his own family's opposition to his efforts.

He smiled broadly at her. "One of these days the whole world is going to be wanting to fly. Your dad and I, along with many others, are going to keep working on this idea until we have a flying machine so big and so safe a dozen people can fly in it at one time."

The expression on her face showed her total skepticism about that possibility.

"We'd better find John and start for the house. Mama won't be happy if we're late for supper," she said. "There's a little pond up the creek where fish bed underneath the tree roots. Let's see if he's up there."

She led the way and they walked along the creek bank in silence. Soon he heard the splashing of water as it fell over the dam in the creek up ahead. Betsy had been right. As they came into sight, there was John fishing in the pond. He had caught several nice trout and had them on a string, dangling in the water beside him.

"You better saddle up, John. Mama has called us to supper," Betsy said.

Her brother looked up and glanced questioningly from Clark to her.

"Oh, this is Clark Dyer, the chap from Georgia who has come up to spend a couple of days with our dad and learn about gliders," she said.

John frowned. "Hello," he said, looking away from Clark. Then as he gathered his fish and got to his feet, he asked, "You do much fishing down in Georgia?"

"Yes. We have some nice rainbow trout and catch a lot of them down there."

They retraced the trail to the road and turned back toward the house without much conversation. Clark wanted to learn about their interests but felt it best to let them lead the conversation as they chose.

Finally, John asked, "Did you make the trip up here by yourself?"

"No. Two of my brothers came with me to my uncle's house in Louisville. I came over here by myself this morning. My brothers wanted to visit with our family some more, and they don't have any interest in gliders anyhow," Clark said.

"I can understand that," John said.

"Do you have any sisters?" Betsy asked.

"Yes, six. Three of them are a lot older than I am, but the other three grew up with me. I also have six brothers."

"Wow! That's a big family. There's just John and me in our family."

When they arrived at the house, Cora had filled the dining table with bowls of food that filled the air with a pleasant aroma.

"The washstand is in there," Betsy said, motioning Clark toward a little room at the side of the kitchen.

Once everyone was ready, they took seats at the table. Edgar asked a blessing and the food was passed around the table. Clark noticed the types of food were very similar to what he was accustomed to having at home. While they ate, their conversation centered on an exchange of information about gliders and the similarities and differences between life in Georgia and Kentucky.

"I guess you have to be pretty self-sufficient to live in those remote mountains, don't you?" asked Edgar.

"Yes, you do," Clark said, "but neighbor helps neighbor whenever special skills or more hands are needed for a job. And we probably are used to making do with a little less than people in better developed areas are. We hardly notice any lack since all of us are in the same situation."

"You must have good schools there. You're a well-spoken lad and you certainly have a lot of creative ideas."

"We have a little one-room school in our neighborhood and have had some very good teachers there. We have an academy in town, but I haven't attended it. I think parents in our area probably take a greater interest in educating their own children than parents do in some other areas. We have always had plenty of books around, some that we owned and some that we borrowed. And when someone goes into Gainesville, they bring back a few newspapers and magazines to pass around the neighborhood."

"No doubt, all of that helps. Betsy loves to read and we're hoping she will decide to become a teacher."

"Daddy, I don't know if you really want me to become a teacher or if you want me to help you keep building better gliders," Betsy said with a teasing smile.

"I would gladly settle for you helping me, but I know you need an occupation of your own that will provide income for you," Edgar said, giving her an affectionate pat on the arm.

"Well, I don't know what I want to do," John piped in, "but it's not teaching or building gliders. Maybe I'll try horse racing or sailing ships."

"It never hurts to try a few occupations while you're making up your mind," Edgar said. "A fellow needs to have satisfaction in his work. Even when life hands you a responsibility to fill that isn't your choice, you can usually develop a hobby to give you fulfillment in pursuit of your dream."

The conversation continued for another hour or two before bedtime arrived. As Clark lay in the comfortable bed Cora had prepared for him, his mind once again turned to his family. He was beginning to feel quite homesick for them, and he wondered if Morena's family had moved away. The thought of that brought sadness to him. When he compared her to Betsy, he knew that Morena's quiet spirit was exactly what he wanted in a wife.

The following day as Clark and Edgar explored different designs for the gliders, it was clear to him that Edgar was quite a philosopher as well as a talented builder. Clark knew he was gaining insights for successful living along with learning techniques that would help him in building his own flying machine one day.

That afternoon as he prepared to leave, he told Edgar, "I can't really express how much I have enjoyed being here. Getting to take a ride on your glider and learning how to build one has been a very special treat. If you ever get a chance, please come down to Georgia and see me and my family."

"I'll do that, Clark. I have enjoyed spending time with you as well. You gave me some good ideas that I plan to incorporate in my gliders. You have a talent for this. Don't let it go to waste."

They shook hands and Clark climbed into his uncle's buggy. He drove away with his head full of thoughts and plans.

16. Love in the Air

The sun was sinking slowly behind the mountain range, spreading orange and gold streaks across the feathery clouds as Clark hurried toward the Owenby home. The dreaded event had happened—Morena's family was planning to move from Choestoe to North Carolina before long to be near relatives who had moved across the state line just before the Indian removal. Clark didn't want Morena to go and was hoping he would get a chance to talk to her privately and find out if she felt the same way about being separated from him.

He had left home in late afternoon to walk over for the visit with the Owenbys. He carried his lantern, expecting it to be dark by the time he made the return trip. As he neared her home, he saw her raking leaves in the front yard. Her younger sisters and brothers were running about playing tag and teasing Morena by running though the pile of leaves she had raked. Just then a burst of wind hit the mound of leaves and scattered most of them across the area she had just cleared.

"Hey, dear friend," Clark said laughingly as he walked up, "you'll never get through raking with all of this going on. Take the kids to the back and play jump rope with them. I'll rake the leaves."

"Hello, Clark. You showed up at the right time. I was about to get a switch to these kids."

"Can you bring me a sack to put the leaves in?" he asked as he took the rake from her. "The wind is going to keep undoing our work if we don't get these leaves out of its path."

"I'll bring one for you," she said, giving him a grateful smile as she turned and called to the children, "Who wants to play jump rope?"

He watched her as she ran toward the barn to get a jump rope. He thought back to the time when he first saw her as she came up the path to their house to ask about taking weaving lessons from Ma. He remembered how he and Jimmy were fascinated with her tanned skin and flowing hair. She now kept her hair wound into a bun and wasn't nearly as tanned as in those younger days. He recalled how pretty she had been the day they attended a fund-raising event for the school when he bought the box lunch she brought. He had enjoyed her company tremendously that day and afterward looked for every opportunity to talk to her.

They had participated together in a number of neighborhood activities in the past couple of years, and Clark had grown to like her even more. He believed that she also liked him a lot, but now with her family's imminent plans to move away there wouldn't be further opportunities for them to get to know each other better.

"This is the biggest sack I could find. Will it do?" Morena asked, returning with a large tow sack.

"Yeah, I can use that," he said. "If you have the kids settled down now, will you hold the top of the sack open so I can stuff the leaves into it?"

As they raked and packed leaves, Clark wove questions into their conversation about what Morena wanted to do in the future.

"Would you rather live here or in North Carolina?" he asked.

"I would like to live here," she said, "but I love being with my family. If I let them move away without me I think I would die of loneliness."

"It wouldn't be easy, I'm sure. I've never had to live away from close family, but we do have some relatives living in Kentucky and Tennessee. When I went up to see them two years ago it was the first time they had seen any of the Georgia family in twenty years. They have missed seeing all of us, of course, but they are happy in their lives there."

"I don't see how you could ever get used to that," Morena said sadly.

"Well, I think when your life fills up with satisfying activities and your circle of family and friends grows, you start to lose that almost constant feeling of sadness about their absence in your daily life. Of course, you continue to feel a stab of pain from time to time as you think of them and miss their presence, but writing letters and visiting whenever you can always help to lessen that."

Clark wondered if he was convincing her that she could stay in Choestoe and still be happy here when her family left. Pa would give him some land on which to build a home for himself and his bride when he married. He had been hoping his relationship with Morena was going to develop to the point where he could expect her to say yes if he proposed marriage. He stopped raking leaves and looked directly at her.

"Do you know how sad I will be if you go away?" he asked.

"I don't want to talk about it," she said. He could see the unhappiness in her eyes before she turned away. "I'm going to see if the kids are okay. Mama will be calling us to supper in a little while."

He finished clearing the yard by himself and went over to the porch and sat down dejectedly on the steps. Mrs. Owenby came out soon and greeted him warmly.

"How are you doing, Clark? I think you came by here at a good time, as I see how the trees are shedding their leaves like rain showers. Thank you for getting the yard cleaned up."

"Yes, ma'am. I'm glad to be here and help out a little. I wanted to come over and visit some with you and the family this evening before you go away. I guess you'll be moving over to North Carolina soon, won't you?"

"We haven't set a date yet, but we're planning to go before winter sets in. Robert is thinking we'll try to be packed and relocated before Christmas. Jonathan moved up there back in the summer and has found a place for us to live."

"It surely will be lonesome here without you all," Clark said.

"Well, why don't you just pack up and come with us?"

"I can't go right now. Pa is counting on me to be his right-hand man at the slash sawmill we built on the creek in Upper Choestoe near where Uncle Cager lives. There's a lot of building going on in that area, and folks want to get finished with their structures by the end of winter so they'll be ready to start planting in the spring."

"This will be the second time we've moved away from Choestoe," Mrs. Owenby said. "We moved the first time to go with family members just before the Indian removal. This time we're hoping to find employment for all of the menfolk in nearby businesses up there. We will enjoy living near the rest of the family, too. Your family has mostly worked at farming since they came here about ten years ago, haven't they?" she asked Clark, then added, "And apparently they've done very well at it."

"Yes, ma'am. I think it's definitely in our blood," Clark said. "While many other men work as miners, or as tradesmen, or go off to war, the Dyer men have mostly stayed here and farmed."

"Well, I know Morena will miss you when we go. Maybe you can come up and visit us sometime."

"I will definitely plan to do that," Clark said. "But I really wish she would move in with one of your relatives and stay here."

"She's very close to the family and would be grieved if she was separated from us," Mrs. Owenby said. "Lord knows we would miss her too if she wasn't with us. She's an industrious young lady and can do more work than I can. She's a little older than you, isn't she, Clark?"

"Yes, ma'am. She's almost three years older than I am, but we really enjoy the times we can spend together."

"I've noticed that you do." She stood up and patted his shoulder. "I must get in the kitchen and get our supper on the table. Robert's been out in the field with the boys all afternoon gathering corn and they are going to come in directly as hungry as bears. You come on and wash up, then tell Morena to get the kids rounded up and ready to eat, too."

Clark followed her past the kitchen to the washstand and quickly washed his face and hands. Smoothing back his hair with his wet hands, he headed out the back door to look for Morena and the younger children.

The boys were playing a game of horseshoes in the barnyard and the girls were jumping rope. When they heard that supper was ready, they raced toward the house leaving Morena and Clark behind to put away the playthings.

"Morena, I want you to write me if you do move away. Your mama has invited me to come up and visit whenever I can. I don't want you to forget me," Clark told her.

"Of course, I won't forget you," she said.

"You promise?"

"I promise."

"Then seal the promise with a kiss."

Clark pulled her to him and lifted her chin. He looked into her soft brown eyes for a moment then placed his lips over hers. He half expected her to pull away, but she melted into his arms. She felt so right there and he didn't want to ever turn her loose.

"Oh, Morena, I really don't want you to go!" he cried. "I love you and I want us to get married."

"Clark, I also think we're a good match for each other but you're too young to get married right now. Besides, Mama needs me to help her out a little bit longer with the kids and the work of moving. Let's wait and see what we think about it next year."

He wanted to push her further about it now, but the resolute manner in which she stated her opinion on the subject convinced him there was little chance he would be able to change her mind. Anyhow, on the positive side he was ecstatic that she had agreed to consider marrying him next year.

While the family sat at the table eating that night, her parents took note of the way Clark and Morena glowed every time their eyes met. In fact, the interaction between the two of them wasn't lost on Morena's teenage sisters, Barbara and Ann, either. They had often talked behind her back that she would probably be an old maid because she never took an interest in any of the young men, and many of those were now becoming wed to other young women in the community. The sisters thought Clark was too young for Morena, but it was quite apparent that he was smitten with her.

By the time the meal was finished and dishes were washed, the night air had turned quite cool. Morena's twelve-year-old twin brothers, John and Joseph, brought in wood and built a fire in the hearth. Everyone sat around the fire while Mrs. Owenby knitted wool socks, Mr. Owenby whittled kindling wood and the children played games. Morena brought in a guitar and handed it to Clark.

"Let's play and sing a few songs," she said. She had a flute that she herself would play, and she passed a fiddle to her father.

"I'm sure we all know 'Barbara Allen.' That's a good one to start with."

"Daddy," she said, "you start it for us."

Mr. Owenby drew his bow across the strings, Clark tuned in on the guitar and Morena on the flute, and soon the house was filled with music and song. They followed up with "Greensleeves," "The Nightingale" and "The Willow Tree."

"I wish someone would write happy songs," Morena said. "All of these have such mournful stories."

"Well, let's do 'Turkey in the Straw,' Mr. Owenby said. "That one, at least, has a light and lively tune to it."

As they began the music again, the youngsters all began dancing and the rest of the family clapped their hands in rhythm.

When the tune was finished, Mr. Owenby said, "We will do a closeout now to send everyone off to bed. What shall it be?"

"Let's sing 'How Firm a Foundation,' Mrs. Owenby said.

Their voices blended beautifully on the old hymn they were accustomed to singing at church. As they finished, Clark stood and handed the guitar to Morena.

"I'm sorry to see the evening come to an end. That was fun." Turning to Mrs. Owenby, he said, "Thanks for the delicious supper."

"If we don't get to see you again before we leave, don't forget to come to see us," Mrs. Owenby said.

"I'm hoping you will keep me up to date with letters telling me how you all are getting along," Clark said, looking at Morena. "I promise I'll write back and give you the news from Choestoe," he added.

He shook hands with Mr. Owenby and asked, "Could I use one of your kindling sticks beside the fireplace to light my lantern?"

"Sure, you can," he said, handing him one of the small sticks.

"You be careful going home, son. Take a pole with you in case you have to fight off a bobcat or something. It always pays a feller to be prepared."

"Yes, sir, it does."

Clark held the stick over the fire in the hearth for a moment until it flamed up, then he drew it across the wick in the lantern. When it blazed, he replaced the globe and adjusted the flame upward. He glanced across the room at the family and gave a last, long look into Morena's eyes.

"I wish you all a safe trip," he said. "Have a good night now."

He walked out into the cool night air and thought all the way home about what the future might hold for them, and for himself.

17. Unsettled Plans

Water hit the paddles on the horizontal wheel with rhythmic slaps at the slash sawmill on Stink Creek. The circular motion of the turning wheel was converted to a back-and-forth motion by a pitman arm that was attached to a saw blade. A movable carriage, which was also water-powered, moved the logs steadily through the saw blade slicing the logs into lumber.

Since the mill had been built, the local men no longer had to cut lumber by hand for their building projects or haul their logs several miles to get them cut at other mills. It was saving hours of labor and transportation for them.

Clark took particular pride in the mill because he had carefully studied other mills in the area to learn how to design this one. His success in getting it operational had raised his standing in the neighborhood. While most of the men knew he had an unusual ability to create innovative ways of doing things, the mill far exceeded anything they had imagined he was capable of designing.

"I guess you're right proud of Clark," Seth Brown said to Elisha as he watched the men lift logs onto the carriage and position them so the saw would cut the size lumber needed for his building.

"I sure am," Elisha said. "I have never seen anybody work as hard as he did to get every part of the mill to work exactly right. When he wasn't able to find parts he needed, he drew and measured and calculated until he got everything to fit so it would operate the way it should. We might not have always appreciated Clark's talents in the past, but this has certainly proved that we're lucky to have him around."

"Yeah," Seth said. "I'm impressed that we don't hear him bragging about what he can do. He just works diligently on whatever the project happens to be. And it's obvious he enjoys anything he makes. I came by here late one evening. It was nearly dark, and there he was, still fitting paddles onto the water wheel. It was the night Grady Oakes was having a corn shucking over at his barn with music and singing. Nearly everybody was going, especially the young people."

Seth laughed. "Maybe we better give Clark a hint that he'll never snag a wife unless he gets out to these shindigs and talks to the girls."

"I imagine he'll find one someday, or else she'll find him," Elisha said.

Clark had decided to go pick up the family's mail at the Choestoe General Store as the sun was rising on this chilly March morning, but when he reached the trail that led to Stink Creek, he opted to go by the mill first to see if everything was running smoothly. As soon as the mill came into sight, he saw the water wheel turning and wagon loads of lumber sitting beside the millhouse. He could hear the whine of the saw as it sliced the lumber. He stopped and enjoyed the sight for a few minutes, and then turned and headed back towards the store. No need to disturb the work, he thought, since everything was running well.

It had only been a month since Clark had gone to North Carolina to visit Morena and her family. Upon returning, he had hinted to Pa that he would like to have a few acres along Stink Creek where he could build a house someday. He was very pleased when Pa told him that he liked Morena and was of the opinion that she would make a good wife for him. He assured him that he would give him some land when he got ready to build.

As Clark arrived at the General Store, he was still pondering how he could get Morena back to Choestoe so he could make plans with her for building their house.

"Good morning," the storekeeper greeted Clark as he entered the warm store. "Have a seat over here by the stove and I'll get your mail for you."

When John returned with the mail, Clark was surprised to see a letter to him from Morena. Afraid there might be bad news since she normally would not be writing so soon after his visit, he quickly opened the letter. It read:

March 9, 1842

Dear Clark,

Mama has been having a lot of pain and is worried about the care she is getting from the midwife here. She wants to come back to Choestoe to have the baby. We will be staying with my Uncle Porter so Mama can be treated by the doctor there. I will send word to you when we get to Uncle Porter's house. We should arrive in the next few weeks.

Affectionately,
Morena

Clark felt uneasy about Mrs. Owenby's condition. She must be seriously worried to leave the rest of the family in North Carolina and come back to North Georgia for treatment. Morena had not said in her letter whether her sisters, two-year-old Eveline and five-year-old Salina, would come with them, but her sisters, Barbara and Ann, were still living at home and likely would be able to run the North Carolina household for a few months during their mother's absence. Jonathan and his wife lived nearby, and Clark felt sure they would help their sisters in managing the four lively young brothers who could sometimes present a challenge. He knew Mr. Owenby well enough to know that he would also find plenty of work for the boys to do if they tried to give the girls a hard time.

Clark chuckled to himself thinking about how Pa used to dream up so many chores for him and his brothers to do when they were rowdy. It was a tactic that unfailingly quieted them.

He wondered how Morena's temporary residence here was going to affect his plans for building a house and getting married.

Spring rains had come generously, and the corn and beans in the fields were about four inches high now. Weeds and grass had tried to overtake the crops, but Pa had put everyone to work chopping out the offenders. Now, rows of green plants striped the brown soil across the valleys and knolls, giving promise of a good harvest to come in the summer.

Morena and her mother had moved in with Porter a couple of months back, and Clark had been able to get over to visit with them several times.

Clark and Elisha sat resting on the back porch in the pleasant late-May sun. They had finished plowing the corn in the big upper field and decided to go fishing for the remainder of the afternoon as soon as they were rested. They heard a horse coming down the road, and the dogs began to bark. Elisha rose and went to see who it was. Frank Brown turned his horse and wagon into the yard as Elisha came around the corner of the house ordering his dogs to be quiet.

"How are you, Frank? And you, Priscilla?" Elisha greeted them.

"We're fine," Frank said. "We stopped for a few minutes at Porter Owenby's house to say hello as we passed by there and learned that Mattie Owenby gave birth last week."

"This is her twelfth child, a daughter," Priscilla said. "She named her Sarah, and despite the poor health Mattie had for months before delivery, the baby is strong and doing well. Mattie has regained her health and is planning to return to North Carolina in a couple of months."

Clark was relieved to hear that the baby had arrived in good health. Now he and Morena could begin making their plans around moving her mother back home. He wanted to talk to

Morena and see if they could set a date for their wedding and reach agreement on where they would live. He was thinking it might be best to rent a house until they could get settled down. The neighbors were always accommodating when there was a need to build someone a house, and he felt sure everyone would pitch in to help build one for him and Morena. He just didn't know if Morena would want to launch into the project right now since she had her hands quite full helping her mother with the new baby and getting them moved back home.

As soon as the Browns went on their way, Clark presented the issue to Elisha.

"Pa, I am going to talk to Morena about this, but I'd like to know what you think. I'm hoping that we can set a date for our wedding sometime this year. I would prefer for it to be this summer, but I will see what she thinks. If it is to take place this summer, do you think we should rent a house to live in, and wait awhile to build our own house?"

"I think you're trying to reach a decision before you have all the facts," Elisha replied. "If she says she wants to wait until this fall or winter to get married, you will have plenty of time to build your own house."

"Pa, I'm strongly hoping she will say she wants to get married this summer."

"Well, if she does, I think you're right that throwing in a construction project while you're getting the wedding plans together would be a mite too much for any bride. Morena has faced a lot during the past few months, and it wouldn't be fair to expect her to do it."

"I want to go over and see her tomorrow. We're caught up with the work in the fields. Do you mind if I go?"

Elisha laughed and slapped him on the back. "Son, you're not going to be able to get your mind on anything else until all of this is settled. Go on over and see her tomorrow. And, by the way, I hope she will agree with you to have a summer wedding."

"Thank you, Pa. I'll be mighty happy if she does."

18. Clark Takes a Bride

Clark walked up to Porter Owenby's front door shortly after sunrise. A heavy dew blanketed the grass and the tender leaves that were growing on spindly bushes beside the porch. As soon as he knocked, he heard the heavy footsteps of Porter coming to answer.

"Good morning, Clark," Porter said as he opened the door. "You're getting an early start today, ain't you? Come in and sit a spell with us."

"Howdy, Porter. We heard the good news that Miz Mattie has had her baby. We're certainly happy to hear that there weren't any complications."

"We were a little anxious about whether everything would be all right," Porter said. "But Dr. Harrison had built up Mattie's system with some herbs he's used successfully with other women, and he was confident she would have a good delivery. We sent for him as soon as she went into labor, but by the time he got here, the baby was already making an appearance. I think Mattie could have delivered without his help, but he always enjoys bringing babies into the world. Since he was responsible for getting her strong again, it was fitting that he actually got to attend the birth."

Hearing Clark's voice, Morena came into the living room carrying her baby sister wrapped snugly in a patchwork quilt. "Good morning, Clark," she said shyly. "Excuse me, Uncle Porter. I want to show little Sarah to Clark."

She held the baby down and Clark gently pulled the quilt aside. Smiling at the baby's sleeping face, he said, "I would say she looks like her big sister, but to tell the truth, I think all babies look alike when they are this young."

Morena laughingly replied, "It's an adorable face for her, but I don't think it would be an attractive look for me at my age."

As she turned to go back to the kitchen with the baby, she added, "Mama is up if you want to come in and say hello to her."

"I'll be along in a few minutes after I visit some with Porter," he said.

"I think they'll be heading home in about a month," Porter told him quietly. "I know Mattie will be glad to get back to her little girls that she had to leave behind when she came down here. The Lord has blessed her to regain good health so fast. She's a strong, praying woman and I believe she has instilled the same qualities in Morena."

Porter turned to Clark. "Am I correct in my understanding that a wedding may be taking place with you and her pretty soon?" he asked.

"Yes, sir, I am hoping so."

"I believe you two will make a fine couple. I know Morena very well since she's been living in our home for the past two months and, of course, I know your reputation in the community, too. I would like to go ahead and tell you now that I wish God's blessings upon you and her."

"Thank you, Porter," Clark said. "Your confidence in us means a lot to me. I am very happy that she's willing to marry me. I have never really been interested in courting any other girl. Now, if you will excuse me, I'm going in the kitchen to talk with her and her mother. We need to make some plans about when we can set our wedding date and where we will live. I'm anxious to begin working on the arrangements for all of it."

Clark rose and shook hands with Porter. As he turned and started to the kitchen, he could feel his cheeks burning with pride at the compliments Porter had paid him and Morena. With both his and Pa's approval having been given for the planned marriage, he felt confident they had made the right decision.

As he entered the kitchen, the baby was crying with soft little mewing sounds and Morena was rocking her back and forth in a cane-bottomed chair. Mattie was sitting in a rocking chair beside the cooking stove, and she waved him over to sit in a chair near her.

"It's good to see you again, Clark," Mattie said with a wide smile. "The Lord has strengthened me and added this fine baby girl to our family. How are you doing? And how are your folks?"

"We're all doing fine, thank you, Miz Owenby. Ma is planning to get over here to see you before you go back to North Carolina."

"I have really missed having her as a neighbor. Only the good Lord knows whether we'll ever get back to Choestoe to live again. Our men folks have been able to find work up there and we've gotten settled in pretty good."

Mattie looked at Morena rocking the crying baby. "Morena, bring the baby over here to me," she said. "It's about time for her to nurse anyway. You can go visit with Clark on the back porch in the sunshine."

Clark was pleased to notice that Morena seemed happy about getting released to spend some time alone with him. She handed the baby to her mother and got her jacket. As she walked out ahead of him, he admired the graceful steps that set her long skirt swaying. Her dark hair was neatly rolled into a loose bun fastened at the nape of her neck. She was altogether a vision of loveliness, he thought.

"I will pull the chairs to the other side of the porch where we'll get some sun and also be sheltered from the cool breeze," Clark said.

As they got comfortably settled, Clark took her hand. "Please say that we can arrange to get married this summer. I don't see any reason for us to wait until later in the year."

"One good reason is that you're not yet twenty years old," she laughed.

"That is not a good reason, Morena. I will be twenty in July, and everybody around here accuses me of acting older than I am anyhow."

"Don't take me seriously in saying that, Clark," she said soothingly. "The fact that there's a few years difference in our ages seems to concern some folks, but I don't foresee that as making any difference in whether we have a happy marriage."

"Do you want us to have the wedding here in Choestoe?" he asked.

"I think it would work out better if we did. My parents can come down and spend a couple of days with Uncle Porter, and we can ask your Ma and Pa to let us have the ceremony in their back yard. Since their household isn't overrun with youngsters like ours, it might be easier for us to set up everything there."

"That would suit me," Clark agreed, "and I don't think Ma and Pa will mind at all if we have it there. Of course, Melinda and Matilda will be overjoyed to have a wedding take place at our house. We'll probably have to rein them in or they will begin making arrangements on a scale fit for royalty. I guess you would like for your Choestoe Church pastor to perform the ceremony?" he asked.

"Yes, I would," Morena said. "Can you go by and talk to him soon?"

"I plan to go and see him next week. Of course, you know he's going to try to convince me to join his church."

"Well, he feels as strongly about his Baptist religion as you do about your Quaker religion," she said.

"Don't worry. We've had that discussion before and we each know where the other stands. Of course, I will never interfere with your attending church there, and I will let him know that. But I love our quiet Quaker meetings. They satisfy a deep need in my soul. It's what I grew up with, and although I won't insist that you attend with me, I would love for you to join me from time to time and see what our services are like."

"I'm sure I will enjoy attending your meetings," Morena said. "But just as you grew up with your church traditions, I grew up with mine and I love them."

"I know you do," Clark said. "Let's never let ourselves be divided because of the differences in the outward trappings of our religions. We're both Bible believers and that's the important thing."

Morena laid her head on his shoulder. "You're a good person, Clark. I am always so comfortable with you."

He turned and wrapped his arms around her. "I can't wait to take you home with me."

She settled happily into his embrace and they wiled away a couple of hours discussing further wedding plans until the sun began sinking behind the mountain and the air started to grow chilly.

"I'm going to head home, Sweetheart. I hope the days will fly past quickly until we are together for keeps," Clark said, giving her a long kiss.

"Me, too," Morena said, as she looked up to him. Her brown eyes were shining brightly and Clark felt his heart would burst with happiness.

Indeed, the days did fly past leading up to their wedding date. Clark had barely procured basic furniture for their rental house, some of which he built himself and some of which was wedding gifts from family members. He had a gold wedding band made for Morena and it was tucked into the pants pocket of his new black suit. Ma had given him a fresh haircut and he felt satisfied that he was ready for the ceremony.

As Clark expected, his sisters had thought of little else for weeks except planning for the big day. They had corralled their friends to help with flower arrangements, singers, musicians and the food to be served. They had counted on the weather to cooperate so everything could take place out of doors. Temporary tables had been set up with planks across sawhorses and covered with their newest and prettiest cloths. Ma had made the traditional stack cake from sun-dried apples. A very large crowd was expected, so the cake was two feet in diameter and twelve layers high.

The musicians arrived early to get their instruments in tune. They arranged their chairs under the shade of a big oak tree near the back porch. They wanted to be sure if a sudden summer shower began they would be able to move quickly to the porch.

As the guests began arriving, Melinda and Matilda met each of them at the front porch. They wanted to make sure when Morena arrived they got her into the house where Clark wouldn't see her until she appeared for the ceremony. It was bad luck for him to see her on their wedding day before the ceremony began, they had been told, and they weren't taking any chances.

The preacher arrived with his wife in their buggy just ahead of Morena and her family. Melinda took the preacher to the back porch while Matilda took Morena into the front bedroom and closed the door.

"Are you nervous?" Matilda asked as she took the veil Morena was carrying and placed it on her head.

"Yes, pretty much so. I hope everything goes right."

"Oh, how pretty you look!" Matilda exclaimed as she stepped back to admire Morena in her long white dress with the veil framing her dark hair and face.

"Where is your bouquet?" Matilda asked.

"Ann has it. Go ask her to bring it in here unless she is keeping the baby. If that's the case, Ma or Barbara can bring it to me."

They could hear the music from the back and conversation and laughter among the guests on the front porch. Clearly, the wedding had put the family and neighbors in a festive mood.

As Ann came through the door, Morena was happy to see that her bouquet had survived the trip without wilting in the July heat. The bottom of the stems of an arrangement of vivid blue delphiniums and powder blue hydrangeas intertwined with Queen Anne's Lace had been wrapped in strips of cloth and placed in a bucket with a little water. Ann removed the bouquet, dried off the water, rewrapped it with a dry cloth and tied a white satin ribbon in a fancy bow.

"Ann, you are so creative. That is just perfect!" Morena said.

"I brought some ripe cherries, too," Ann said, taking a small jar from her bag. "Rub some on your lips, but be very careful not to let the juice get on your dress. I picked daisies for Barbara, Melinda and Matilda to carry," she added as she took three more bouquets from the bucket and tied them with white ribbon. I think these will look good with the blue dresses they're wearing, don't you?"

"Oh, yes, I do," Morena said.

"I have boutonnieres for Clark, Lige and Jimmy. See." She held up a white rose. "I have a blue ribbon to tie on Clark's rose so it will match your bouquet. I also made them for the preacher and for Pa and Mr. Elisha." Ann held up two corsages. "See what I made for Ma and Miz Elizabeth."

"Melinda, can you take the boutonnieres and pin them on the fellows?" Ann asked.

"Yes, I'll do that," Melinda said, coming over and taking the roses from Ann. She paused and gave Morena a hug before heading out the door. "I'm so happy for you and Clark. It has been a barrel of fun planning your wedding."

"I'm going out to see if everything is in place for the ceremony to start right away," Ann said. "The preacher is here, and the fellows and girls should be about ready to line up. I'll come back for you as soon as I check it out."

Alone in the room, Morena walked back and forth wondering if Clark was nervous, too. Probably not, she thought. He had looked forward to their marriage for months, and when he set his mind to do something he moved steadily ahead to get it accomplished with no fear of failure. She hoped she could learn to be more like that.

Ann opened the door and crossed the room quickly. Taking Morena by the arm, she said, "Time to go, big sister. Pa is waiting to walk you down the aisle."

They went into the kitchen and stood at the door beside her father, who looked a little anxious

"This is good practice for you, Pa," Ann said, hoping to ease his tension. "By the time you give both Morena and Barbara away, you'll be an expert when it's your turn to give me away."

"I might help you and Barbara elope so as to avoid this whole ruckus," Robert said, but he was able to muster a smile for her.

Ann stepped out on the porch and saw that the preacher was standing in place and the musicians were ready with their instruments. She looked to see if the young men and girls were ready to walk down together to the stand set up to serve as a temporary altar. They were in place and she nodded to the musicians to begin playing.

As the violin softly played the opening strain of "Greensleeves," the wedding party began their slow march forward. Morena took her father's arm and they moved toward the doorway, stopping short of it so Clark would not be able to see her yet. His grandparents had been seated, and so had Morena's mother. When the bridesmaids and groomsmen were in place, the musicians began playing "Wedding March." Robert and Morena traded brief smiles and started on the path to the altar.

Clark thought he had never seen a bride more lovely than Morena as she came toward him. It seemed surreal that in the next few minutes they would exchange vows and be man and wife. Her sisters had thought her too reserved to catch the fancy of a beau, but to him her quietness was her most attractive characteristic.

Robert and Morena stopped in front of the preacher, and he solemnly intoned, "Who giveth this woman in holy matrimony?"

Robert replied, "I do." He took her hand from his arm and placed it in Clark's hand. Clark gave her hand a gentle squeeze and hoped it helped her to relax some.

The preacher gave the traditional introductory readings that he always used for weddings. Then, following the question, "Does anyone here have objections to this marriage?" and no responses, he asked Clark and Morena to join both their hands and repeat the vows after him. They could hear Morena's mother sniffing and knew she was quietly crying as her daughter became a wife and left their household to make one of her own.

The vows finished, the minister closed with, "I now pronounce you man and wife. Clark, you may kiss your bride."

As soon as they kissed, he turned them to face the group and said, "I present to you Mr. and Mrs. Clark Dyer."

The musicians burst forth playing "Coronation March" and family and neighbors gathered around them with hugs and handshakes. The young people began circle dances, line dances and informal jigs and reels. Finally, running out of steam they began gathering around the table for the elegant meal prepared by the family and neighbors.

By the time Clark and Morena could get away to head to their house, they were totally depleted of energy. But as soon as they got into their house and collapsed on the bed, they heard the shivaree begin. There were loud greetings to them, including banging, hollering, and serenading. This went on for more than an hour before the serenaders were finally quieted down by a thunderstorm that came up.

"Well, Mrs. Dyer, what do you say we try to get some sleep now," Clark asked Morena, observing the exhaustion in her eyes.

She gratefully sank into his waiting arms. "It was wonderful to have everyone here for the big celebration, but I'm happy to have it over. I look forward to the days ahead when there's just the two of us together."

Suddenly, a large clap of thunder shuddered the little house as if saying their old way of life had passed and a new one was beginning.

19. Loss of a Best Friend

Clark sat beside his workbench in the lean-to attached to the barn, shoulders slumped, eyes downcast. He wanted to do some work on his flying machine, but he felt loss so heavily he couldn't even think of where to begin. The family had laid Pa to rest in the family burial plot the day before, and there was no one else on earth who took the kind of interest in his outrageous idea that Pa had.

Pa didn't understand the concept of how a machine could fly, he thought, but he had enough confidence in me to believe that I knew what I was doing.

He had appreciated Pa's support, but only now did he fully comprehend the value of that support.

An hour passed and still he couldn't make himself pick up a tool to begin work. He heard Morena calling to him from the house. He rose slowly and started down the hill. Morena walked toward him with quick steps.

"Are you okay?" she asked anxiously. "Supper's ready. Come on and eat."

As they walked into the house, three-year-old Jasper came running to him and wrapped his arms around his leg. Sixteen-month-old John began squealing, "Papa, Papa!" Clark felt some of the tension ease in his shoulders as the love of his young family surrounded him.

Morena smiled at him. "You'd think they hadn't seen you in a week the way they are carrying on."

He and Morena would be celebrating their fifth anniversary soon. With the help of their neighbors, they had built a very comfortable home for themselves. He had even piped water into the house from the big spring, and Morena was so excited to have at hand all the water she needed without having to carry it from the spring in a bucket.

"Are you feeling a little better?" Morena asked as she searched his face. "I know you will miss Pa—all of us will."

"Yes, it's a loss for the whole family," Clark said, "but I believe I have been closer to him than they have. His pride in my work and the fellowship we shared when he came to the shop and sat talking with me about what I was doing, about how you and the kids were doing, giving me advice on how to deal with any problem I had, I will miss that more than I can ever express. I'm thankful I had these twenty-five years of my life with him, but I wish I could have had at least twenty-five more."

"Time will help ease your grieving for him, but I know it will never completely go away," Morena said. "Ma has always told me that as long as a deceased person lives in your heart, they are never really gone. Remember that when you're feeling the loss and talk to Pa in your heart."

Clark gratefully wrapped his arms around her, realizing how blessed he was to have an understanding wife. He picked up John, Morena took Jasper by the hand, and they headed to the kitchen.

"Eat! Eat!" Jasper shouted, upon seeing food on the table. Pulling Morena's hand, he danced and waved his free hand wildly. John joined his brother in the chant, laughing and bouncing in Clark's arms.

"Hey, boys!" Clark warned. "Let's calm down and be little gentlemen."

As the youngsters settled down to eat, Clark asked, "Did you know that Eli joined the Army to go fight in the War with Mexico?"

"Yes, I did. Sallie came to visit your Ma yesterday and told her about it. She said Andrew joined, too."

"Now, why would they want to do a thing like that, I wonder," said Clark.

"They seem to think the pay is better than what they can make working around here. Sallie told your Ma that she wasn't sure she would be able to tend the farm with only Elisha and Thomas to help her."

"The crops are mostly made now," Clark said. "We will all chip in and help her harvest as she needs us. Maybe Eli will send a little money home for hiring some farm help, too." He felt irritated that Eli would leave Sallie in such a predicament. He knew he certainly wouldn't think of putting Morena in a dilemma like that.

Turning his thoughts elsewhere, he remembered the books Uncle Joe had sent him by Jimmy when he was in Gainesville doing some trading a few weeks ago.

"Where did you put the books Uncle Joe sent me?" he asked Morena.

"They're in a box beside the bed," she said. "I saw that he had put a note on top for you to look at a newspaper clipping he enclosed for you about a man in Belgrade who tried to fly an ornithopter. I see other clippings in there, too. You must have kindled his interest in flying."

"When I was just a kid Pa took me with him to Gainesville and we spent the night with Uncle Joe and Aunt Margaret. You should see his library! I wanted to stay a week or two and just read. Uncle Joe packed a bag of books for me to bring home. Ever since then, he has made a habit of saving newspaper stories and books on history and science to send me by anyone from here he meets up with."

"He's a special person to do something like that," Morena said. "I hope I get to spend time with him and Margaret someday and get to know them better."

"We'll try to make that happen. I feel so encouraged right now just thinking about him."

When they finished supper, Clark took the two youngsters in the living room to play while Morena cleared the table and washed the dishes.

"Jasper, how would you like for us to make an airplane?" Clark asked.

"No! I want a buggy!" Jasper said.

"Okay, we'll make a buggy, just a little one that you can ride Little Buster in. Let's go to the workshop and get something to make it with."

Clark got a tin pan and a large spoon and put them in the floor for John to play with. He called to Morena, "Keep an eye on John while I take Jasper with me to the shop to look for something to build a buggy with."

He lit a lantern and took Jasper's hand. They stepped outside and headed up the hill toward the shop. The katydids were beginning to tune up for their evening serenade, and their raucous screeching echoed across the mountains.

"Papa, what makes that noise?" Jasper asked.

"It's a whole bunch of katydids talking to each other," Clark said.

"What are they saying?"

"They're saying they want to play."

"I would play but I can't see them."

"We'll build a buggy at the barn and you can play with it. That will be more fun that any ol' katydid, won't it?"

Suddenly, they heard something crashing through the bushes at the edge of the pasture. Clark pulled Jasper to his side and whispered sternly, "Stand still and be quiet. There's something in the woods!"

The light from the lantern shone only a few feet from them and made seeing into the darkness impossible. Clark hoped the noise was a deer and not a bear or panther. He realized now that he shouldn't have come out without his gun. He slowly lowered the lantern to the ground and stepped away from it pulling Jasper with him until they were beyond the circle of light. He picked him up and held him against his chest. "Sh-h-h," he whispered in his ear. Thankfully, Jasper seemed to understand that they were in danger and didn't make a sound.

As Clark kept inching toward the barn in the darkness, he could hear footsteps of the animal moving toward the lantern light. God, help us make it to the barn, he prayed silently. The barn had always seemed near to the house, but now it felt like it was a mile away.

Now his eyes were getting more accustomed to the darkness, and he saw the shadow of the creature begin moving into the ring of light. It was a huge bear! It approached the lantern and took a swipe at it with its front paw. The lantern went careening down the hill with the bear following. Clark ran toward the barn, stumbling in the darkness. Jasper began to cry, but the bear was focused on the tumbling lantern and didn't stop to investigate the new sound.

Finally reaching the barn door, Clark felt for the latch and yanked the bolt back. He pulled the door open with one hand and with the other thrust a screaming Jasper inside. As soon as the door was closed and fastened, Clark picked him up again and stroked the back of his head.

"You don't have to be afraid now, little buddy. We outran that ol' bear and he can't get in this barn. We'll just wait right here until he goes away."

Jasper slowly stopped crying and began to relax. "Where is the lantern?" he asked.

"I'm afraid that rascal tore it up. We'll have to wait until tomorrow to make you a buggy. Is that all right?"

Jasper nodded. "Papa, will you shoot that bear tomorrow?"

"I will if I can find him. He may have headed across the mountain looking for something to eat. He travels lots of miles to find his food every day."

Clark opened the barn door and peered through the darkness. He didn't hear anything stirring and decided it was safe to start back to the house.

"Jasper, can you hold to my hand as we walk back to the house?"

"Uh-huh," Jasper said.

"If I hear or see anything I will pick you up and run with you again, okay?"

"Okay."

Clark could see the shadow of their house at the bottom of the hill and made his path in that direction. The katydids were still cheeping noisily, unmindful that there had been a calamity in their midst.

As Clark and Jasper entered the front door, Morena came from the bedroom with John in her arms. "Couldn't you find anything to make a buggy with at the barn?" she asked.

"Tell her what happened, Jasper."

"Mama, a bear came out of the woods and hit our lantern. He tore it up and we ran to the barn so he wouldn't get us, too. I was scared, so Papa carried me."

"Oh, my goodness!" Morena said. "Where did he go?"

"I don't know," Clark said, "but he was so taken with the lantern he tore it all to pieces before he went on his way. It's a wonder he didn't set the grass afire with the flame and kerosene bouncing around inside the lantern as it rolled down the hill. I will try to find it tomorrow and see if it can be fixed."

Clark was thankful the incident had ended without there being property or bodily injury. They had no neighbors within sight or sound and an accident of any kind at night would be especially disastrous.

"Let's all kneel and thank the Lord for his protection, especially for tonight but also for every day," he said to Morena.

They all knelt beside the settee and lifted grateful prayers to God. Even the toddlers were still and quiet in the solemnness of the moment.

As he rose from prayer and headed to bed, Clark realized that the day which had started for him with such despair over the loss of Pa was now coming to a close with a new assurance that God would be beside him to meet whatever he had to face in the days ahead. He would build a flying machine, and Pa would smile down on him from heaven for not giving up on a dream that was so special to him.

He put his arms around Morena. "And I thank God for giving you to me. You are definitely a fitting helpmate for me."

"And you for me," she replied. "Your tender spirit is perfect for me."

20. More Than Gold

The newspapers being brought back to Choestoe by farmers who traveled to Gainesville lately were becoming filled with stories about the discovery of gold in California. Clark was thinking how with gold mining about petered out at the local Coosa and Duke's Creek mines, and even greatly declining at the Dahlonega mines, it was to be expected that some folks would travel all the way across the country to try their luck at mining in California.

The newspaper article he was reading said, "Letters the miners are sending back to their families are causing 'gold fever' to build to a high pitch around here. They tell of gathering large amounts of easily accessible gold—in some cases, thousands of dollars worth each day. They are claiming that even ordinary prospectors are averaging finding gold worth ten to fifteen times the daily wage of a regular laborer here in Georgia. They say a person can work for six months in the goldfields out there and find the equivalent of six years' wages that they can make here at home."

As Clark sat reading the newspaper, Morena was nearby nursing their newest son, Henderson, while Jasper pulled John around the room in a little wagon that Clark had made for him.

"What do you say we load up these kids and move to California, Honey?" Clark asked her.

"Clark Dyer! You know full well that a stick of dynamite couldn't blast you out of Choestoe," Morena retorted.

He threw his head back and laughed heartily. "You're right about that. I didn't even hire on at the mines around here when so many others were doing it," he said. "Morena, I guess I'm just a born and bred farmer, since the only jobs that interest me are farming and building things."

"Well, you're making a good living for us with your work," Morena said, "so it's a good thing you enjoy what you do and that we can enjoy staying right here."

"I went by to see Ma this afternoon," Clark said. "Now that the Mexican War is over, Eli and Andrew have been discharged and are back home with Sallie. Ma said Eli's war injury seems to be doing very poorly and he's not able to do any work. But it is a big help to Sallie and the rest of the family having Andrew back to lend a hand on the farm."

"Is your Ma doing okay?" Morena asked.

"She's managing fine. With Matilda and Lump still living with her and now Jimmy and Eliza are building their house nearby, the four of them take care of anything she needs. I had a good time with her sharing recollections about some of Pa's escapades in bygone years. That seemed to get her in a very good mood. I know it surely was delightful for me!"

"I'm glad the two of you got to visit together some," Morena said. "I'm beginning to be lonesome for a visit with my mama. My little brother, Robert, turned three last November, and I think it's time for Mama to plan a trip down here to visit us. It's hard to believe that Sarah is already six. And Eveline and Salina are nine and twelve. I don't want all of them to grow up without knowing who I am."

"Write her and tell her to come as soon as spring gets here," Clark said. "I want to put in a lot of time working on my flying machine between now and planting time. I've been thinking about some different ways I can curve the wings of the flyer and make them tilt so they can capture the force of the wind. This would give a navigator the ability to turn the machine in any direction he wants to go. It would be an amazing accomplishment if I could succeed in building wings that would do that."

Morena smiled at his excitement as he talked about the flying machine.

"I have no doubt that you can do it, my dear," she said. "Everybody knows you're a genius in figuring things out. It's a gift from God and you're supposed to exercise it."

"I don't have to force myself to work at constructing this thing. If I had enough time and money to buy everything I need for building and trying out the ideas I have, I would be out there in the shop working sixteen hours a day."

"I know you would, and I'm glad you have an outlet for expressing your creativity. There's no telling what you might come up with someday," Morena said.

"I don't talk much about the flying machine to people because usually they think it's just a silly idea," Clark said, "but I read what the newspapers are saying. There are people all over the world right now trying to figure out how they can make a machine that will fly. It's going to happen. I have no doubt about that."

"Well, it will certainly be an exciting day when it does," Morena said. "As delayed as we are in getting news here in the mountains, we probably won't hear about it for a month or two after everybody else has heard it."

The dogs began barking and they heard a man's voice call out, "Hello. It's Zeb McCleary here."

"That's ol' Zeb. He has a liquor still back in the hollow beside a little creek," Clark said. "I'll go out and see what's on his mind."

"Howdy, Zeb," Clark greeted him as he walked onto the porch. "Where are you headed?"

"I'm needing somebody to help me load some whiskey and bring it to my house. Can you lend a hand?"

"Sure, I'll be glad to help you out, Zeb. Let me get my coat," Clark said.

"He's needing some help. I'll be gone for a few hours," he told Morena as he came back in. He pulled on his coat, donned his hat, and gave her a quick kiss on the forehead.

"You boys be good for your mama while I'm gone," he said to Jasper and John.

"Clark, please be careful out there," Morena said. "Take your gun with you. One of those crazy drunkards might try to attack you and Zeb."

"Oh, you worry too much," he replied, but he reached over and took his gun from the rack and went out the door with it under his arm.

Zeb noticed the gun and said, "I guess you must have heard that there's been some robberies at the stills here of late."

"Yes, I did. Whiskey is getting to be the new gold around here," Clark said. "I understand there's a wagonload or two taken to Atlanta about every day."

"We know how to make good whiskey here and the Atlanta folks trust us to keep it clean and safe. Word gets around when anybody starts selling tainted liquor. Let somebody get sick, or worse, they die from drinking poison booze, and everybody hears about it."

"Well, I know that a fellow can get much more value for his corn if he first converts it to whiskey" Clark said. "After all, one horse can haul ten times more value on its back in whiskey than it can in corn."

"Right," Zeb said. "Since I got in the business, it has meant a difference in how well I can take care of my family."

As they approached the still they scanned the nearby trees for any unusual movement. Seeing nothing, they went over to the shed, which was filled with jugs of whiskey.

Zeb's horse was tied to a nearby tree, and he already had it hitched to the wagon. He brought it over to the shed and they began loading the jugs.

They had the wagon about half loaded when the horse suddenly lifted its head and stiffened its ears as it looked into the woods. Zeb grabbed his gun and said quietly to Clark, "Get your gun. The horse hears something or somebody coming. Step behind that tree and cover me until I know if it's friend or foe."

Zeb threw a blanket over the wagon, closed the door to the shed and sat down on a stump, keeping his eyes fixed on the trees. His gun was pointed in the direction of the horse's gaze.

A few minutes passed and Zeb heard slow, measured steps approaching. He had his finger on the trigger and his body was tensed. He had never had to shoot anyone and he hoped this was not about to be an event that would change his record.

Out of the woods a tall, thin man appeared carrying a knapsack on his back. He didn't look to be armed, but Zeb was taking no chances. "What's your business here, stranger?" he asked.

"My name is George Featherstonehaugh and I'm traversing the mountain area making a record of the minerals and geography here. I smelled the aroma from your still and came to see what kind of whiskey you make."

Zeb studied him intently. He had a pleasant expression on his face and he looked him squarely in the eye. His accent clearly was not Appalachian. Zeb rose, leaned his gun against his leg and reached his hand toward the stranger for a shake.

Featherstonehaugh grasped his hand with a smile. "I'm pleased to make your acquaintance Mister..."

"My name is Zeb McCleary."

"Do you mind if I call you Zeb?"

"No, not at all. Can I pour you a drink, George?"

"Yes, I would be obliged for a drink of your whiskey."

Zeb picked up a pint jar and stepped over to the wagon. He raised the corner of the blanket and brought out a jug of whiskey. He poured some in the jar and handed it to George.

George took the jar, shook it vigorously and watched the bubbles. "I am told that large bubbles with a short duration indicate a higher alcohol content. If there are smaller bubbles that disappear more slowly it indicates lower alcohol content. Do you know if that is true?"

Zeb said, "I think it's a pretty good test. Also, the folk test for safe quality of moonshine by pouring a little of it into a spoon and setting it on fire. The theory is that safe liquor burns with a blue flame, but a tainted liquor burns with a yellow flame. I know my still and the ingredients I use will reliably make pure whiskey, so I don't test it after it's made."

Clark came from behind the tree, and Zeb introduced him to George. "We're having some trouble with thieves around here and you can see that we're staying armed because of that," Zeb said.

"What is your recipe for the whiskey you're making?" George asked.

"Our main ingredient is corn mash. Immigrants from Ireland brought a recipe with them for what they called the 'water of life.' The whiskey isn't aged like I understand they do it in Europe."

George took a small swallow from the cup. "The flavor is really good," he said.

"Thank you. I'm glad you like it," Zeb said.

"I'm just coming into your community after traveling through several other places recording the minerals, gems and rocks of each area. It appears that gold mining has declined here in the Southern Appalachian Mountains. Are there any mines still being worked here?" George asked.

"Not many. Close by we have the Duke's Creek and Coosa Mines which are about depleted. Farther away, they're still mining in Auraria and Dahlonega, but they don't get much gold there either. All the big miners have left, and just the small-timers remain doing a little panning," Zeb told him.

"You have a lot of minerals throughout this area in addition to the gems and rocks, but I can see that getting large equipment in here to handle any mining would be very difficult."

"That it is," Zeb said. "We do our farming and make a little whiskey on the side to raise enough money for paying our taxes and buying things we can't produce. We have to be content with that. But I don't have any complaints. I'm a pretty satisfied man."

"That's worth a lot to a man. Different occupations suit different folks. Myself, I love traveling around the country like I do, recording what natural resources are in each place," George said. "Well, I guess I'd better be getting along now. How far is it to a lodging place?"

"Are you heading for Robertstown or Blairsville?"

"I'm going to Robertstown."

"If you take the trail through Low Gap, it's about five miles to the inn at Robertstown. The trail's not too steep. You can probably make it in about three hours. Good people own the inn there and they will feed you a big, tasty meal you'll be talking about for a long time."

George extended his hand to Zeb. "It was a pleasure to meet you fellows. I wish you success in your whiskey venture."

Zeb shook his hand. "I enjoyed meeting you, too, George. Have a safe trip."

Zeb and Clark watched him as he quickly crossed the creek on the foot log and headed up the trail.

"Well, that was certainly a welcome surprise," Clark said. "He didn't know how primed we were to shoot somebody, did he?"

"No. He seemed like the kind of man who never expects to run into any kind of danger. I've heard about him. The government has him traveling around, writing down everything about the land. There's no telling what they think they can use all that information for. It's a waste of time if you ask me," Zeb said.

"Let's finish loading this whiskey and get it safely to your house before the sun goes down," Clark said. He knew Morena was going to become worried about him if he was gone too long. No doubt, she was also longing for his help in looking after the boys so she could cook supper without interruption. But even when he wasn't there, he knew she was completely capable of keeping the three of them in line by herself.

"Yeah. No need to give our wives a fit of worrying about us," Zeb said. "Mine frets over what might have happened to me when I don't show up at the time she expects me to be there. She knows about the dangers involved with making whiskey and would like for me to do something else. I really enjoy making the stuff and it doesn't make money for me."

The two men snuffed out the fire under the still and went down the mountain with the wagonload of liquid gold.

21. Conflict Within the Nation

W hen Clark arrived in Blairsville, he tied his horse to the hitching post in front of the courthouse. Rain from the night before had left the dirt roads muddy, and the old mat at the courthouse door was grimy from the scraping of many boots worn by the men entering the building. The sun had come out and the Liars' Bench alongside the building was occupied by several men talking and spitting tobacco juice in long streams.

"Howdy, Clark," one of the men said as he approached.

"Howdy, Seth. I hope you're doing well this morning," Clark replied, as he nodded a greeting to Henry, Paul and Thomas, who were sitting on the bench with Seth.

"Well, I guess I'm doing okay," Seth said. "You got time to talk with us awhile about the news that's making the rounds?"

"Sure. I've been working around home the past couple of weeks and don't know about anything that's going on. Has something terrible happened in the county?"

"The consequences of it haven't hit Union County yet, but it's a federal issue that's going to affect the whole country before it's over."

Clark pulled a chair over and placed it at a right angle to the bench and sat down.

"If the news is about a federal issue, I don't think a fellow has to do much guessing to say what the issue is. Congress should have known when they passed the Kansas-Nebraska Act allowing those two territories to choose between slavery and free soil that fighting was going to erupt over which it would be," he said.

"You're right in presuming it is the slave issue," Henry said, "but now the fight has moved beyond Congress. A pro-slavery legislature was elected in Kansas because 6,300 ballots were cast in a region that only had 3,000 voters. Now abolitionists in New England and other parts of the North have formed Emigrant Aid Societies and they are sending anti-slavery activists into Kansas where they can vote to keep it free. In Georgia and Alabama, similar societies are sending in settlers who will vote in defense of slavery. Several people have lost their lives in battles between the opposing sides."

"Well, I just don't see how their dispute could entangle us folks here in the mountains," Thomas cut in. "Hardly any of us have slaves and the few who do treat them so well they don't seem to have any desire to be free."

"You're probably right that some of them would be fearful if left on their own without a master to assign their work and provide for their needs. But I believe for the ones who understand what freedom means in allowing them to go where they please, to own property and make their living in whatever way they choose, it would be a joyful day for them if they were granted their independence," Clark said.

"It seems clear to me that most of the men fighting to keep slavery are doing it for the protection of a guaranteed labor force to operate their businesses," Thomas said. "But if they were to set their slaves free and offer to pay fair wages for the work they're doing, he could probably expect to keep most of them."

"I think you're right, Thomas. You make so much sense we should send you to Congress," Seth said.

The men laughed boisterously at the idea of Thomas, who had recently had his eightieth birthday, getting elected as a Congressman and moving to Washington.

Clark said, "I'm afraid we have a crew up there in Washington making decisions on matters they don't have a speck of understanding about. And, on the other side, we don't have much understanding of our responsibility to discuss our concerns with our representatives so they can see our point of view."

"Well, the problem, as I see it," Paul said, "are the splinter groups who care nothing about what's good for the nation; they just enjoy the big fight that's going on. They keep folks stirred up by making inflammatory remarks about what the opposing side is saying. They are telling them that the other side is conniving to obstruct them from reaching their goals. We would do well to focus our efforts on stopping those groups so we could have some sensible discussions among ourselves about how to handle this kind of dispute."

"If things keep going the way they're now headed, we will wind up having a civil war," Seth said.

Clark was struck by the similarities between the way this dispute was proceeding and his memories of how the Indian removal issue had been dealt with years ago by the federal government when he was a teenager. He felt that Seth was right, the nation was headed for a serious rift—he couldn't bring himself to even think of a civil war taking place in his country.

"Fellows, I have to get my property taxes paid and head back home," he said rising from his chair. "I pray this thing can be resolved peaceably. Otherwise, I fear that Seth is right, our country is in danger of breaking apart over it."

After Clark paid his property taxes, he left the courthouse, gave the men on the bench outside a farewell wave, and went across the road to Harvey's Dry Goods store. He picked up several items that Morena had asked him to get, then decided to buy some small toys for the children. For the boys he got a bag of bright colored marbles, a cup-and-ball game, a checkerboard, and a chalk board with chalk. But what could he get for three-year-old Cynthia? Then he spied a little basket with five clay eggs that he knew immediately she would love. She enjoyed watching Morena gather eggs from the nests in the barn every day, and this would help keep her from getting underfoot as Morena worked. He finished his shopping by filling a little bag with assorted candies from the large candy jars at the counter.

Harvey took a pencil from behind his ear and carefully recorded on a scrap of paper the price of each item that Clark had placed on the counter. "That will be eight dollars and forty-five cents," he said.

Clark pulled money from his coat pocket and handed it to him.

"How are your wife and kids doing?" Harvey asked.

"They are all doing very well. Thanks for asking"

"You're getting quite a big family now, aren't you?"

"We have six children—all boys but one. I don't mind if we get six more. We really enjoy them. We don't have any close neighbors, so these youngsters keep the place lively."

Harvey smiled broadly and held out his hand for Clark to shake. "It's good to see a strong family. Keep up the good work. Much obliged for your business."

"You're welcome," Clark said. "I'm glad to have your store here locally. You keep a good supply of goods so I don't have to travel so far when I need something. Have a good day."

Riding back home, Clark thought over what the men at the courthouse had said about the dissention over the slave issue. He worried what it would mean to Uncle Joe and Aunt Margaret if the federal government took away the state's right to allow its residents to own slaves. Uncle Joe's six-hundred-acre farm was worked by three men slaves and their five sons, and Aunt Margaret had three women slaves and their four daughters who did the cooking, cleaning and laundry for the household. He wondered whether all of them would stay if Uncle Joe asked them to do so in return for a fair salary. If they didn't, what would he and Aunt Margaret do? They were getting old and it would be impossible for them to run the house and farm by themselves.

I'm glad our families don't own slaves, he thought. If the country goes to war, that will be hard enough to handle without the additional burden of trying to deal with fifteen slaves like Uncle Joe would have to do.

By the time he got back home Clark had decided he would not burden Morena with the news about a threat of war. She had her hands full with managing the household along with their growing family. He was pleased to have such a good wife who had the ability to handle the demands of their busy life. They had bought land from time to time and now owned four hundred acres consisting of a hundred acres in cultivation and pasture with the rest in woodland. He was thankful that he had achieved so much at his young age of thirty-three. He knew he couldn't have done it without her help and encouragement.

When Clark arrived home and rode his horse to the stable, Jasper and John were nailing beaver hides to the side of the barn for drying. They expected to make a fair amount of money from selling the dry hides to the tannery.

As soon as their father dismounted, they were at his side to see what he had brought from town.

"You'll have to wait till we get to the house to see what I have," he told them. "We'll let the whole family see the things at the same time."

The eager boys removed the bridle and hung it on a peg. They brought hay and filled the horse's manger, talking all the while between themselves about what Papa might have in the saddle bag.

"I hope he brought us some marbles. I've been wishing hard for some," John said.

Clark came around the barn where they were talking in time to hear what John said. He was pleased that he had decided to get the marbles. His boys were good and didn't complain when they had to help with the many chores on the farm. They deserved to get a little gift now and then.

"Come over to the shop with me," he said to them. "I want to press some corn shucks I boiled for making wings for my model flying machine."

The boys exchanged quick looks. It was obvious that they weren't anxious to watch boiled corn shucks being pressed—not with a saddle bag of treats waiting to be unpacked. But they tagged along behind him nevertheless.

Clark pulled a pot from under the workbench in the shop. It was half full of wet shucks. He lifted out a clump and shook off the water. He placed the shucks on the workbench and spread them evenly. He lifted out more clumps until he emptied the pot and spread all the shucks on the workbench. Then he placed lumber on top to hold the shucks flat.

"I'll leave the lumber on top till the shucks are almost dry, then I'll remove it so they can finish drying without being held down. That will make a sheet of material smooth as cloth and the shucks will stick together to make the piece airtight. The material will be easy to cut, so I can make the wings whatever size I need for the flying machine," Clark told them.

Clark stepped over to a shelf on the other side of the shop. He took out a frame for the body of a model which he had made from cured river canes and held it up for them to see.

"I'll have to decide just what size wings I need for a model this size and the best place to attach them," he said, turning the frame this way and that as he mulled over the options.

He was smiling in pleasure over his project, but he could see that the boys were anxious for him to finish so they could get to the house and see what he had brought.

"Let's go, fellows. Mama is probably wondering what's keeping us up here."

"Mama knows exactly what's keeping us," Jasper said with a laugh. "She knows you can hardly bear to let a day pass without spending some time up here in the shop working on something."

"Your mama is a good woman, boys. I hope you both will be as fortunate as I was in finding a wonderful wife."

"Papa, I think your flying machine is going to be the best one that's ever been built," John said.

Clark gave him a hug. "Thanks, son. Keep pulling for me. I have dreams that I will eventually fly over our fields in a machine that will look a lot like this model. I believe it will happen some-day."

22. Surprise Election Results

The approaching Presidential election was the center of discussion wherever folks gathered during the hot summer days of 1860. For the past ten years, national conflict between the pro- and anti-slavery groups had raged furiously. The mountain people thought William Seward would be the Republican nominee for President this time because of his favorable position on slavery and state's rights. But at the Republican Convention the delegates judged him too weak to win in some key northern states in the General Election and selected instead the Convention's second choice, Abraham Lincoln.

"I think selecting Lincoln as the nominee is going to cost us the General Election in November," Caleb told the men lounging in the Choestoe General Store playing checkers.

"It's primarily the southern states with a large population of slaves that will suffer if they are prohibited from owning slaves. They're not going to vote for Lincoln, you can be sure of that."

Caleb rose from his chair and began pacing the floor, waving his arms as he continued pressing his views. "You see, we have all those western states with very few slave owners, so it's not just the slave issue. They do, however, have an issue with the federal government telling them what they can and cannot do in their own territories. Now, you put these two groups together and they have enough voting power to whip the northern states in this election."

"I wish I had as much confidence in the outcome as you do, Caleb," Ethan said, "but a victory by Lincoln will almost guarantee secession by the Southern states. They have been openly threatening secession from the Union since the beginning of the year if Republicans happened to gain the White House."

"He's right about the South's strong opposition to Lincoln," Herman said. "They're afraid of the anti-slavery policies of the Republican Party, and they're going to fight like the devil to defeat Lincoln. Don't think for a minute that they have forgotten the Lincoln-Douglas debates in the Senate race two years ago. I myself clearly remember how Douglas defended the principle that each state should have the right to decide whether it would become free or slave, and all the while Lincoln argued fiercely against every policy that would allow slavery to continue or spread within the Union."

"I still don't see how Lincoln can win the election," Caleb insisted. "Voters are bound to stand with Douglas because they want a president who understands how ridiculous it is to have outsiders telling states what they can and cannot do. I think that will trump everything else."

Ethan shook his head. "I think it's more than that, Caleb. The northerners are divided on the issue of whether it is humane to withhold freedom from any citizen. Many of them don't feel a state should have a right to decide who does or does not have freedom. For them, it is not a plan for telling people what they can do but for allowing every person to have the freedom guaranteed by the Constitution."

The men grew quiet and sat solemnly, each thinking about the difficult situation their nation faced, wondering what the future held for all of them. The game of checkers no longer held their interest and they began making excuses about needing to go home and take care of chores. As they filed from the store, their shoulders were stooped and their eyes stared straight ahead, revealing just how worried and dismayed they were about the situation.

———————

The crops had been harvested and the days were growing shorter. Clark had been able to complete work on his new wing design for the flying machine by stealing away to his workshop every time he could spare an hour or two from farm work. As he sat scrutinizing his model on the worktable before him, his expression reflected his pleasure at having reproduced the curve of the maple seed in forming the wings. He had taken the corn shucks he boiled and dried earlier and moistened them enough to bend over a form he had made in the shape of the maple seed. Now dried in that shape, they were ready to be attached. If tomorrow brought steady winds blowing across the field, he would test the model to see if it operated the way he expected.

He was tying the wings to the body of the model when he heard the hoof beats of a horse coming up the trail from the creek. He rose from his stool, looked down the pathway and saw Jim Lance on his palomino. A faint smile crossed his face. Jim was his cousin who lived a few miles away, and he was Clark's biggest supporter in the work of building the flying machine.

Clark stepped out the door to try to get Jim's attention and let him know that he was at the shop, but Jim had already passed the house and was headed up the path to the shop.

"Howdy!" Clark greeted him. "You knew where to find me, didn't you?"

Jim laughed heartily and slapped him on the shoulder. "Now, Cousin, just about anybody knows where to find you if you're not in the field. How's everything going for you?"

"I can't complain. What about you?"

"No complaining by me either."

"Well, come inside and let me show you what my latest idea is to get better lift for the flying machine."

Jim ducked his head as he went through the shop door. He was half-a-foot taller than Clark and was in the habit of ducking whenever he walked under any doorway. The shop door was tall enough to accommodate his height, but barely. He picked up the model and examined the new wings Clark was attaching.

"You curved these in a different way I see."

"That's so they'll catch more wind underneath and lift the machine quicker and higher. Do you think it will work?"

"It makes sense to me. I wish we had enough wind to try it out this evening."

"Maybe you can come over the first day we're getting good wind and see how it performs."

"Yeah. I'll plan to do that."

"I guess you're like the rest of us, weary over all of the discussions and arguments people are having about who to vote for in the upcoming election," Clark said.

"Oh, my goodness! I can hardly believe the passel of notions they all have about this thing, even within my immediate family. I think we essentially agree on the point that it wouldn't matter to us if the slaves were given their freedom, because none of us have slaves anyhow and we know it would be the charitable thing to do for them. But as we consider the effect it would have on the large plantation owners, we begin to see the problems it would create. Those owners buy and sell slaves like we do livestock. And there's a line of thinking by some that the slaves aren't smart enough to make a living for themselves with no owner to manage things for them if they are set free."

"Well, it's plain to see that the big farms wouldn't be able to operate the way they have done for more than a century if they didn't have the slaves," Clark said. "I've thought about that. But if they set them free then hired them back, they ought to be able to keep operating their farms in some fashion."

"They should," Jim agreed. "I have been reading the newspapers and they say the views are split nationwide about which of the candidates would be best able govern the country and heal the rift that's been growing with every passing year. You know, it really is a dilemma."

"It's typical of the kind of arguments the political parties seem to get into every election," Clark said. "This time I actually think it has more to do with the idea of not letting the federal government tell the states how they can run things than anything else."

"It could be," Jim said. "There is not much need for us to get too disturbed about it, I guess. We can't solve the problem for them."

"Well, here we are taking the same direction everyone else does and talking about the election," Clark said. "It has got all of us worried for sure. We'll be lucky if we get through this one without serious problems."

"We're lucky to be living here in the mountains away from most of it. I had better be heading toward home now," Jim said as he started out the door. "I'll be back the first windy day that comes to watch you test the new wings."

"Yeah, do that, Jim. I think I've got a good curve formed on them now. I will finish attaching them before I head to the house for supper in a little while. Stay for supper with us if you will," Clark invited him.

"No, Jane is expecting me home soon. She'll be worried if I don't show up, so I'd better get on my way now."

Clark watched Jim ride down the trail. His horse pranced along hastily as though he knew they were headed for the barn. Jim sat upright in the saddle, whistling and looking as if he didn't have a care in the world. Clark felt a sudden twinge of envy for such a person, knowing that he wasn't driven to try to accomplish a feat that consumed so much time and effort. Yet, Clark knew a disposition like that would not fit with his psyche. He was serious minded and had to have a challenging project in process to feel gratified.

He turned and stepped back into the shop. I wouldn't trade places with Jim for the world. This is the life I was meant to live, he thought.

Election Day finally arrived and voters went to the polls in droves all across the country, wanting to have a voice in who would lead the nation during such troubled times. When word reached Choestoe that Lincoln had won, Clark went into town to see if any newspapers were available to confirm the news.

Entering Harvey's Dry Goods store, he saw a *Charleston Daily Courier* on the front counter bearing the headline "Honest Abe Elected." He picked it up and began reading:

"Lincoln captured slightly less than 40 percent of the vote, but he won a majority in the Electoral College, with 180 electoral votes, by sweeping the North (with the exception of New Jersey, which he split with Douglas) and also winning the Pacific Coast states of California and Oregon. Douglas won nearly 30 percent of the vote but won only Missouri's 12 electoral votes. Breckinridge, with 18 percent of the national vote, garnered 72 electoral votes, winning most of the states in the South as well as Delaware and Maryland. Bell, who won 12.6 percent of the vote, secured 39 electoral votes by winning Kentucky, Tennessee, and Virginia."

"Do you think you will be getting any other newspapers about the election?" he asked Harvey.

"I doubt it. I was lucky to get this one. A man passed through from South Carolina and had a few with him. I told him that we needed to have one here for the local folks to see and he let me buy it from him."

Clark looked back at the paper. "I see that Lincoln didn't win any votes in the southern states except Virginia, and he only got one percent of the total vote there. I think this leaves the South in a critical position."

"His election is going to cause a terrible backlash in the southern states," Harvey said. "The guy from South Carolina said they are already planning conventions at the local level to select delegates to a state convention where they will decide what their response will be to the election. He says there is no doubt that they will vote to secede from the Union, and not only South Carolina but Mississippi, Florida and Alabama, too."

"You don't think Georgia will secede, do you?" Clark asked him.

"I wouldn't be surprised," Harvey said. "I think all of the cotton-growing states will pull out. What other choice is there? They can't run their farms without the slaves."

"I certainly hope there won't be a rush to decide about seceding. I believe we should make every effort to develop some way to function within this system that we find ourselves in due to the election results. We need to remain in the Union," Clark said.

"It seems like when an idea gets ahold of people it's hard to shake it from them, especially when they're holding onto the less painful of two choices," Harvey said. "In this case, they think seceding is less painful than trying to work out some kind of compromise, but the North is not going to sit quietly by while the South continues to own slaves. I'm telling you, this is going to lead to war."

"That's a fearful thought," Clark said quietly, "but we've been heading in this direction for a long time." He rose and put on his hat.

"Thanks for letting me read your paper," he said. "We all need to pray earnestly about this problem. I certainly will be doing that."

"Me, too, Clark. You take care, hear?"

All the way home he felt the weight of the nation's problems on his back. He prayed and asked for God's guidance. Yet his doleful spirit was at one with the dreary winter day.

23. War Affects Choestoe

It was a beautiful June day as the Dyer family gathered to bid farewell to their matriarch. She had pressed on for fourteen years as a widow, growing frail and quiet in the last couple of years and finally becoming bedfast.

Sadness blanketed the crowd of family and neighbors as they sat solemnly before the homemade pine box that held her thin lifeless body. When the congregation finished the hymn "Nearer, My God, to Thee," the minister rose and stood behind the pulpit to speak.

"Elizabeth Clark Dyer was a strong woman of faith," he began. "She was a faithful wife to Bluford Elisha Dyer, Jr. for forty years, up until his death in 1847. They raised thirteen fine children and most of them remain here in Choestoe to this day. They have become valuable members of our community.

"Elizabeth cared for all of her sick and needy neighbors and was a dedicated member of our church. We will sorely miss her presence here, but one day we will join her in Heaven where there will be no more parting, no more farewells to be said to those we love."

Morena reached for Clark's hand and shifted her body against his leg, attempting to find a more comfortable position on the wooden bench. Their ninth child was due in a month, and Morena tried to straighten her dress over her protruding abdomen. Two-year-old Rena had fallen asleep in her lap and was limiting her ability to move. Five-year-old Robert was snuggled against Clark on the other side, and the three next older children sat beside them. Their three teenaged boys sat on the bench behind them. Clark thought how quickly the years had passed since his marriage and how soon it seemed the house had filled with children.

"Let us go forward from this place today carrying in our thoughts the remembrance of the life Elizabeth lived before us. May her life serve as an example for us to follow," the minister said.

A wail came from Matilda sitting on the front bench. Her husband wrapped his arm around her shoulder and leaned his head against hers. Sallie moved over closer to her and took her hand. Clark knew the loss of Elizabeth would weigh heaviest on Matilda. She was the youngest daughter and had been especially attached to her mother. It brought back the memory of the intense loss he had felt when Pa died. Sometimes there is a special bond between family members that brings excruciating pain when it is broken, he thought. He would make an extra effort to sustain Matilda over the coming months, but he knew that she would have to travel the sad path of grief on her own to a great extent.

The minister gave the benediction and the congregation filed out of the church to begin the procession behind the pall bearers transporting the casket to the cemetery on a nearby hill.

"Morena," Clark said as he took Rena from her, "I think it would be advisable for you to let Jasper drive you home in the carriage with Rena and Robert right now."

"Yes, I think so," Morena said. "I will take Cynthia home with us, too. She's not going to want to stay here with just you and the boys, and she is such a good little helper with Rena anyway. Now that I've gotten so large I don't know what I would do without her."

They quietly got into the carriage and Jasper turned the horses toward home. He and Emmaline Lance, who was also eighteen, were starting to show interest in each other, and Morena thought he looked somewhat sad at having to take her home instead of going to his grandmother's home where he would probably get to talk to Emma for a while. The neighbors would have a meal spread for everyone at her house when they returned from the cemetery.

"I'm sorry you're having to take me home and miss getting to visit with everyone, Jasper. Maybe you would like to go on over to Grandma's house after you get us home?"

"I think I will if you don't care. Emma looks especially pretty in that green dress today, don't you think so?"

"Yes, she does.

"Mama, I want her to know that I really like her so she won't get interested in some other fellow. I think she likes me, too."

"Well, I don't know why she wouldn't like you, son. You are mine and Papa's pride and joy."

Morena thought how fast the years had passed. Here was Jasper almost grown and getting serious about a young lady! We have a house full of kids, she reflected, but the possibility of giving up one of them is extremely upsetting.

Cynthia was keeping Rena and Robert entertained in the back seat of the carriage by doing the motions to the song she was singing in her high sweet voice:

"The wise man built his house upon a rock. The wise man built his house upon a rock. The wise man built his house upon a rock, and the rain came tumbling down. Oh, the rain came down and the flood came up. The rain came down and the flood came up. The rain came down and the flood came up, and the house on the rock stood firm."

As she finished the second verse with "And the house on the sand went splash," she clapped her hands together and the children laughed and clapped with her.

Jasper drove the horses near the porch and stopped them. He quickly alighted and ran to the other side of the carriage to help Morena down. Cynthia then handed Rena to Jasper and climbed down to help Robert out. They all went inside, and the house felt empty to them without the rest of the family present.

"There's food on the stove, Cynthia," Morena said as she took Rena from Jasper. "You can fix plates for you and Robert.

"If you want to eat with us, Jasper, she can fix a plate for you, too."

I think I'll go on over to Grandma's house. I should get there by the time everybody else from church does. I will leave the carriage and Blaze here and ride Spirit over there. Will you be okay by yourself with these kids?"

"Oh, yes. I will be fine. You and Papa tell everyone how much I regret not being able to be there."

"Mama, do you want me to fix a plate for you, too?" Cynthia called from the kitchen.

"Yes, Honey. I will feed Rena from my plate. I believe she's about ready to take a nap. Robert, come let me help you change into some play clothes before you eat. I don't want you to drop food on your best shirt and trousers. Cynthia, please put an apron over your dress. You look so pretty in it. I want you to keep it clean."

Morena decided to stay in her own dress. It had been her Sunday dress through the last three pregnancies and she felt she was entitled to get a new one if there was to be a tenth pregnancy in her future.

As Jasper headed to his grandmother's house, he met William Townsend with his mother, Sallie, and sister, Betsy, on their way from the cemetery. Just behind their carriage was Sallie's daughter, Polly, with her husband and two young children.

"Is your mama doing okay?" Sallie called to Jasper. "I noticed that you left with her right after the funeral was over."

"Yes, Ma'am. She was just tired and so was the baby. She said that she'll be okay when she gets some rest."

When they arrived at Elizabeth's house, the porch was over-flowing with people. Jasper scanned the crowd to see if he could spot Emma anywhere. To his dismay, he saw George Henson leaning against the side of the house grinning at her. She didn't look comfortable at his presence, and Jasper decided he would go over and break up the conversation before it could get started.

"Howdy, Emma," he said as he walked up. He gave a pre-emptive nod to George. "Can I get some food for you?" he asked Emma.

"Why, yes. That would be kind of you."

"Get me some, too, while you're at it," George sneered.

Jasper turned toward the porch and said back over his shoulder to him, "You don't look like you have a broken arm."

When he got back with plates for him and Emma, George was still standing there.

"Do you want to sit on the log over there to eat, Emma?"

"Yes, that will be good."

"George, you'd better get inside to the table while there's still some food left," Jasper said. "It doesn't look like anybody's going to bring you a plate."

"Smart Alec," George hissed back at him as he walked away.

Clark appeared with his other four sons in time to hear George's remark. "What was that all about?" he asked Jasper.

"George was hanging around making Emma feel uncomfortable and got mad when I brought food for her," he said.

"He strikes me as a trouble maker. Don't egg him on."

"I didn't really say anything out of the way to him. He's just mad because Emma came with me."

"We are going to get something to eat and then head home to check on your mama.

"Emma, you enjoy your lunch and try to keep Jasper out of trouble."

"Yes, sir. I have to go home soon, too. Maybe George will get too busy eating to try to cause trouble. Tell Ms. Morena hello for me."

As they sat eating, Emma said hesitatingly to Jasper, "I think your Papa is right about George being a trouble maker. I overheard some fellows who were visiting my brothers talking about how he is always making snide remarks to people."

"Well, I would say he had better be careful about making snide remarks around your brother Jim. Jim can probably whip any guy in the community with one hand tied behind his back," Jasper said.

"I don't think you have to be concerned about him when he's alone. He likes to have some other bullies around so they can gang up on a fellow who is by himself," Emma said.

Jasper smiled at her. "I hope I can read into that comment some encouragement that you're concerned about me."

She blushed and dropped her eyes, trying to recover her composure.

Jasper reached for her hand. "Emma, I'm glad you care about my welfare. You're a very special girl and I would like to claim you as *my* girl. Dare I hope for that?"

She looked up at him for a quick moment then back down. She said, barely above a whisper, "I would like that."

Jasper grabbed her other hand, pulled her up and gave her a long hug. "I'm a very lucky man. I love you, Emma."

"I have to go now. Mama will be looking for me," she said.

"I will walk over there with you. My horse is tied to the fence on that side of the house."

He had the feeling that many of the people who saw them walking across the yard together were taking note of it and drawing the conclusion there was a romance blooming between them. Good, he thought, I want them to know that she belongs to me.

As they reached the other side of the house, there was a group of men talking with her father. They were keeping their voices low out of respect for the bereaved family, but a keen interest in the discussion was reflected in their faces. Seth Thomas was saying to them, "Company B, to be called the Choestoe Guards, will be formed from Union County men to help make up the 23rd Georgia Volunteer Infantry Regiment. If you want to join, the recruiters will be at the courthouse next Wednesday."

There was a hum of conversation among the men about how losing their menfolk to the armed forces was going to affect their farms and businesses, as well as the community at large.

"But we have to help fight this war. It's going to take every man we can spare to make up a force big enough to defeat the Union Army," Seth said.

"I'm going to join," William Townsend said. A number of other voices chimed in, "Me, too."

Jasper started feeling a strong urge to speak up and say that he would join, but he felt it wouldn't be right to do so until he talked with his father. He looked over at Emma waiting in the carriage and could see from her furrowed brow that she recognized the discussion involved something which foretold a threat to their future. He wanted to go over and try to ease her anxiety, but that would have to wait until he reached a decision about what he was going to do.

He went across the yard to the fence and untied his horse. He waved goodbye to Emma as he turned and headed down the trail toward home.

He was distressed at this turn of events. The whole neighborhood knew about the battle at Fort Sumter in Charleston Harbor, South Carolina, two weeks ago. They knew that although the Confederacy won that battle, it was due to several favorable factors which wouldn't be present in their future battles. The main factor was the advantage the Confederacy had of the fort being surrounded by water, which enabled them to hold back the

Union forces from getting food and supplies to their troops in the fort.

When he arrived home, Clark was sitting on the front steps. Jasper put the horse in the barn and came back to sit beside him.

"Papa, did you hear about the Choestoe Guards being formed from Union County men to help make up the 23rd Georgia Volunteer Infantry Regiment?"

"Yes, they were talking about it before you got to your grandma's house today. Our country has reached the place I feared we would, Jasper, and now we are in the position of having to take up arms and place our lives on the line. It should not have turned out like this. Reasonable leaders should have been able to put together a plan to solve the slavery problem."

"Papa, I feel like I should join the Choestoe Guards and help defend the South. I'm young and strong, and they need me."

"I understand how you feel, son. I don't doubt that I would be thinking the same thing if I were your age. But being your dad and two decades older, I bear the weight of experience and the dread of what could happen to my first-born son on the battle field. I will leave the decision to you about enlisting, and I will support you in whichever choice you make. You know, of course, your mother's heart will break to see you go. Let's pray about it before we go in the house. Then, you can sleep on it and decide tomorrow what you want to do. Let your mama have tonight to rest without this weighing on her heart."

They bowed their heads and Clark prayed briefly and earnestly, "Lord, we're facing a tough time as a nation and only you know what the future holds. We ask for your spirit to be upon us to lead us in making right decisions. Especially, I ask for you to guide Jasper as he chooses whether to join the army. Let your will be done and give us the strength and courage we need to face whatever comes. I humbly ask this in Jesus' name. Amen."

Morena looked at the two of them as they came into the house and perceived that there was something serious on their minds.

She presumed it had to do with the war, but had no inkling that it was an issue involving her very own household.

Jasper went to bed early to avoid having to make light conversation with the family at this time when his heart and mind were filled with the serious choice facing him.

He was fully aware that army life would be rough. He had heard Eli Townsend tell about the battles that took place when he fought in the Mexican War, how men of all ages were injured and killed, some of them falling dead at his feet.

Jasper had never been away from home and he knew he would miss his family terribly. But he felt it was his duty to go and help defend his state's right to govern itself without having its policies dictated at the national level.

Mama and Papa would certainly miss him and the help they were used to getting from him in running the farm. Life would be harder for them, and they would worry a lot about him.

And there was Emma. Would she wait for him until the war was over?

Sleep was a long time coming as his body tossed and turned, much as his mind was doing.

Part 3

1864 – 1891

They shall mount up
with wings as eagles.
Isaiah 40:31

24. Emergency at Home

It seemed to Clark that the turmoil in the nation had been fomenting for half of his life. The eleven Southern states that seceded from the Union had formed the Confederate States of America, which was considered an illegal government by the United States government, bringing the two factions to a rapidly escalating war.

Jasper had enlisted in the Choestoe Guards, along with several other local young men, and the letters they were getting from him said the number of Confederate soldiers was significantly lower than the Union's. Jasper said he was also noticing in the skirmishes that the Confederacy had considerably less equipment and what they had was mostly inferior to the Union's. Although it was apparent from his letters that the conditions at the camps and in the field were quite bleak, he never failed to include in his letters an assurance that "I'm doing fine. Don't worry about me. This war will be over before long, and I'll be back home with you. I love all of you and miss you really bad."

Clark rose early on July seventh when he heard Morena already stirring in the kitchen. While that would have been normal in past months, it had not been true for the past few weeks. She wasn't feeling well as the time for delivery of the baby drew near. He dressed quickly and went to the kitchen to help her. As soon as he saw her, he knew he needed to get the doctor without delay. She was bent over with one hand on her abdomen and her eyes were filled with tears.

"Honey, go lie down! I will fix something for the kids to eat, then I'll go get Doc Ensley. Do you feel like you will be okay that long, or do I need to go for him now? John and Cindy can fix breakfast for everyone if you think I should go now."

"I'm hurting pretty bad, not sure the baby is going to wait until Doc Ensley can get here."

"Okay, then I will send John to get him so I can stay with you."

Clark went to rouse John and get him headed for the doctor.

"We need you to hurry, son. Your mama needs help right away," he told him.

He went in the kitchen to see what had been left over from last night's supper that John could eat as he traveled. He was relieved to see a baked sweet potato and a piece of cornbread in the cupboard. He wrapped them in a cloth and handed them to John.

"Eat this as you go. Ride Blaze and make him travel as fast as you can. Tell Doc your mother needs his help as soon as possible."

Clark went back in the boys' room and gently shook Henderson's shoulder.

"Wake up, son. I need you to run down to Frank and Matilda's house and get Matilda to come as quick as she can to help your mama. If Eliza can come, bring her, too. I'll have your breakfast ready when you get back."

He went to Morena to see if her pain had lessened any. Her eyes were closed and he decided it was best to leave her resting while he went out back to start a fire under the iron pot. He was happy to see a stack of dry wood nearby that Morena had ready to use for boiling water on washday. He would be able to get water heated by the time the doctor arrived.

He grabbed two buckets and went to the creek and filled them and began filling the pot. He made four trips to haul enough water to fill it. As soon as the fire was blazing beneath the pot, he hurried back to the house to see about Morena. She was awake when he went in.

"How are you doing?" he asked her.

"The pains are close together, Clark. I hope the doctor can get here in time to deliver the baby."

He held her hand and stroked her forehead. "Relax as much as you can, Honey. Henderson has gone to get Matilda and Eliza, too, if she can also come. We can help get the baby here without the doctor if we have to. Just relax as much as you can."

"There's a stack of towels on the chair in the girls' room. Bring them in here," Morena told him. "There are some sheets on the shelf over the clothes rack in the closet. Bring them, too. It's a good thing this is happening in July and not January with bitter cold weather like we had when Henderson was born."

After he had gathered up the sheets and towels, Clark went to the smokehouse and took down a ham hanging from the rafter. He cut a hefty piece from it and returned to the kitchen where he sliced it to fill a large skillet. He added wood to the stove and placed the skillet on the stove eye. He looked at the grandfather clock in the hallway and calculated John should be at the doctor's house by now. His hope was that Henderson was already on his way back with Matilda and Eliza. He was praying silently: Lord, please keep the baby from coming before help gets here.

"Papa, where is Mama?" Cynthia asked as she came into the kitchen.

"Mama's not feeling well, Cindy. She's in bed. I'm starting to cook breakfast. Will you set the table for me?"

"Papa, you can't make biscuits and gravy, can you?"

"I will try unless your Aunt Matilda gets here in time to make them. I'll let you help me and maybe together we can do it. As soon as you get the table set, go to the springhouse and get a basket of eggs for me."

Clark got Morena's mixing bowl and scooped flour into it and was about to add lard when Henderson came through the door with Matilda and Eliza.

"Here. Let me do that," Matilda said, bustling into the kitchen.

Clark gratefully stepped out of her way and replied, "My! Am I glad to see you, Sis! I doubt the kids would have eaten any biscuits and gravy I would have made."

Matilda laughed at him. "Go check on Morena and tell her I am here. I will come in there after I get food on the table unless she needs me sooner."

Hearing the voices, the rest of the children awoke and came to see what was going on.

Eliza scooped up Rena and headed back to the girls' room with her. "Come on, little sweetie. Let's get you dressed for breakfast."

Rena started to cry. "Mama! Mama!"

Clark was rubbing Morena's back and, upon hearing Rena call for her, Morena said, "Tell Eliza to bring her to me for a few minutes."

They placed Rena beside her on the bed, and she stopped crying and cooed at her mother. Then another labor pain hit Morena and she said, "Eliza, take her to Cindy. She can play with her and keep her distracted, I think."

Morena's pain subsided after a while and the children went outside to play. The house became quiet and Clark sat beside the bed, holding Morena's hand and listening to the rhythmic ticking of the grandfather clock. Lord, speed the doctor to us, he prayed again and again. The sight of his wife's weary countenance concerned him.

Another pain came and perspiration trickled down her face. Please, Lord, bring the doctor now. As the prayer went up silently from his heart, he heard hoofbeats in the yard. He jumped up and went to the door. John and Doc Ensley were dismounting.

"Doc, I'm so glad you're here. I hope you have something that will ease my wife's pain," Clark said.

"So she hasn't birthed the baby yet? I'm glad for that. Let me take a look at her."

"Howdy, Mrs. Dyer," he said as he reached her bedside. "Let's get you positioned just right to help this baby come. I'm sure you're more than ready for that to happen."

He asked for Matilda to come in and help get the sheets and towels arranged beneath her and pillows under her shoulders and head. As he opened his bag and began removing his instruments, a powerful pain hit Morena. She no longer held back but pushed down with all her might.

"That's right, Mrs. Dyer. Push hard!" Dr. Ensley urged.

Suddenly, the baby's head appeared and the doctor gently took hold of it. With another attack of pain, Morena pushed and the doctor pulled. Their joint effort was successful and a new baby boy entered the world.

The doctor suctioned the baby's nose and throat, and it gave a thin cry.

"Congratulations," he said to Clark. "He's a fine-looking boy."

He handed the baby to Matilda for her to bathe in the warm pan of water she had brought in.

He completed his care of Morena, gave her a few ounces of laudanum, and then took the now clean and snugly wrapped baby from Matilda.

"Here he is, Mrs. Dyer," he said, handing the baby to her. "I think he may be a little premature, judging from his size."

"Yes, he is little," Morena said, "but a couple of my other children were this size and they thrived right from the start. I trust that it will be the same for this one."

Dr. Ensley pulled a notebook from his bag. "Do you have a name for him?" he asked Clark and Morena.

They both nodded affirmatively.

"His name is Johnson Benjamin Dyer," Clark said.

"How many births has Mrs. Dyer had before this?"

"Eight."

The doctor finished making his notes about the birth, put his notebook and equipment in his bag, and stood to leave.

"By the way, are you still working on building a flying machine, Clark?"

"Yes. I give it as much time as I can spare. Now that the corn is laid by, I hope I can have some extra hours to work on it from now to harvest time."

"I guess you get a lot of ribbing about your labors on this unusual idea, don't you?"

"There's some of that, but not a lot. I think people in the neighborhood know me well enough to understand that I will work on the machine until I get it to fly, or die trying. It's just the way I am when I get my mind set on something. About the only time I get joshing about it is when the subject comes up in the presence of strangers, and around men who don't read and keep up with what's going on in the world. The whole idea of flying is inconceivable to them."

"I have to confess that it's completely beyond my understanding as to how a machine could fly through the air like a bird," Dr. Ensley said. "But there have also been other inventions in the last several years that accomplished things I never thought could be done: steam engines powering ships and trains, electric lights replacing gas street lights, messages sent instantly across the country by way of Morse code. So even if I don't understand how you'll do it, I cast my vote for your success in actually designing a flying ship. I believe you have the intelligence and enough persistence to do it."

"Well, that's mighty nice of you to say that, Doc. I appreciate your confidence."

Dr. Ensley shook hands with Clark and started out the door.

"Send for me if your wife has any problems. I think she's going to be just fine, and the baby, too."

Matilda came out of the bedroom and closed the door behind her.

"I have everything cleaned up and Morena and the baby are asleep. I will go back home if you think you all can take care of everything. Do you want me to come back tomorrow and help out?"

"No, no. We've sent for Morena's mother and she is coming to stay a couple of days and look after things. Thank you for coming today. I don't know what we would have done without you."

"I'm glad to help you out, brother. Morena has helped me so many times I could never repay her. I hope she and the baby will both do well. It's sad not to have Ma here for the arrival of another grandchild, isn't it? She loved being present for the entry of the grandbabies."

"I think of her and Pa nearly every day. I regret that the younger children won't have the pleasure of knowing them like all of us did. They were really special people."

"Yes, they were."

"Eliza, are you going home now?" she called to her sister-in-law in the kitchen.

"I'm going to wash all the linens while there's a fire going under the wash pot. By the time I'm finished, Mrs. Owenby will probably be here to help Morena. I can also fix supper for everyone here before I go," Eliza said.

"Send one of the kids to get me if you need me," Matilda told Clark as she started out the door.

"Thank you. I will do that," he said.

With things settled down, he decided to go to the shop and tinker with the flying machine.

"I'll be up at my shop awhile. If you need any help, just call me, Eliza."

He looked in on Morena and the baby, and they were still resting peacefully. He quietly closed the door and ambled up the hill to the shop.

Entering the shop, he went straight to the shelf that held his books and papers. He reached for the record book in which he had been entering information ever since he was a teen about his experiments in attempting to get a machine to fly. He made a sketch of his current machine with its new wings and did a separate drawing with details of the wing design.

There was not much hope for any significant winds at this time of year, so he began to ponder how he could produce power for his machine without having to rely on the wind. A wind-up spring like those used to propel toys would probably work for the model if he could find one that was strong enough. But one substantial enough to be effective for his purpose would increase the weight of the machine, which would further hinder it from getting enough lift to fly.

The stress of the day had depleted his energy. He leaned back against the wall and closed his eyes. A short nap would renew his strength. He could assess his options for a power source on another day.

25. Wings of an Eagle

Morena had sized up little Johnson appropriately on the day of his birth, she thought, as she watched him chasing the older children around the room. His chubby little cheeks were rosy and his blue eyes sparkled with vitality as he strived to keep up with his brothers and sisters.

Jasper was home from the army following discharge after one and a half years of duty in the Confederate Infantry. He had become disabled from complications of measles and was weak and pale. Morena and Clark felt very blessed that their son had not died from the disease because quite a few of his fellow servicemen had succumbed to it. Morena felt that she could nurse this son back to good health as she had done in the case of Johnson and the other children.

They sat beside the fireplace on a cold January day and Jasper quietly sipped from the cup of hot chicken soup Morena dipped for him from the big black pot hanging from the hook over the fire. She gently rubbed his back and shoulders.

"Eat as much as you can, son. We've got to get some meat back on your bones. And don't be thinking and fretting about the army. You did as much as you could and those who remain will have to finish the job."

"I know, Mama. But this is not how I wanted it to end."

"Of course, it's not. But you have you to whatever life hands you and make the best of it. Happiness and satisfaction only come from within as you give your finest efforts to every opportunity that is presented to you. Your Papa would have been very proud to serve in the infantry with you, but he had me and the children here that he was responsible to provide for and protect. Yet, he is being able to serve in the Home Guard and help the whole neighborhood as he joins with the rest of the men who didn't get to go to the battlefield."

"As soon as I get able, Mama, I'm going to join the Home Guard, too."

"That will be good, Jasper. But my prayer is that the war will be over by the time you get well."

"Mama, I don't see how we're going to win this war. Every attack by the Yankees kills a lot of our men. They have destroyed so many of our buildings and so much of our equipment. Our troops don't have enough to eat, and injuries and sickness are really widespread. I hope it's just my weak condition causing me to see our prospects as being so bleak, but I'm really worried about how this is going to end."

"Don't think about it now, honey. Every one of us will do what we can to help the cause and we'll pray for the Lord to help us through everything we have to face."

Morena saw his eyelids drooping as the warm soup and crackling fireplace relaxed him. She pulled the wool blanket on back of his chair around his shoulders.

"Go lie down for a while. I'll call you after you get some rest."

She held his arm to steady him and led him to the bed. She removed his shoes and wrapped his feet snuggly in the blanket. She pulled up a quilt from the foot of the bed to cover him and was happy to see him drifting into a peaceful sleep.

After he had slept a couple of hours, Jasper opened his eyes and looked around the room, wondering for a moment where he was. The comfort of the warm bed and soft pillow brought a faint smile to his lips. Then realizing where he was, he thought, I will get well again now that I don't spend my days and nights shivering in the cold. And now I have Mama's good, freshly cooked food to eat every day, which will make me burly in no time. Then a sense of guilt assailed him as he thought about his fellow soldiers who were still enduring the discomforts on the battlefield. God, be with them, he prayed earnestly.

He sat up and pushed the cover back. Locating his shoes, he slipped into them and went to the kitchen. He heard Mama and Cindy talking quietly in the next room as they spun wool into yarn. Rena was rolling some kind of toy across the floor and laughing with the baby. They all looked up when he came to the door.

"Are you feeling any better?" Morena asked.

"I sure am. Where is everybody else?"

"John and Henderson went with your Papa to take a load of firewood to Widow Collins' house. Her boys are in the army and she and the girls were getting low on wood. It looks like we're going to get a spell of bad weather by the end of this week. Mancil, Fate and Robert are at school now, but they will be home in a little while."

"When have you seen the Lance family? I can't wait to see Emma."

"We saw them at church last Sunday. Emma asked about you. Maybe you'll be strong enough by next Sunday to go to church with us. I hope the weather won't get so bad that services have to be called off this week."

"If services are canceled, I will try to go to her house next week and see her. I hope she will be happy to see me. I got a letter from her about once a month the whole time I was gone."

"I hear Papa and the boys coming back from making the wood delivery," Morena said. "Now, don't let your brothers talk you into doing anything outside today. You need to stay indoors where you won't get chilled. I know you want to get well, and to do that you will have to take care of yourself for a while."

"I will, Mama. It's easy to do right now because I'm so weak."

John and Henderson came inside with their faces red from the cold wind buffeting them as they had loaded and unloaded the wood, and it had continued to batter them as they rode home on the wagon. The sight of Jasper so pale and skinny stopped them from their usual teasing and joking. They had heard about the heavy toll the diseases were taking on the infantrymen, and they feared for their brother's life. Jasper guessed what they were thinking.

"Come on, fellows. You'll make me go into shock if you treat me like a delicate flower. Pull up a chair and tell me what's going on in the community."

They shed their coats, hats and gloves and sat down.

"There's not a whole lot going on with the war messing up everybody's life," John said. "Nothing's being built, we're not having any celebrations. Nobody's taking trips."

"We're not even taking goods to Gainesville to sell now because people are afraid to get out," Henderson said. "And a lot of people don't have money to spend anyhow."

Clark came in and heard the gloomy reports Jasper was getting from his brothers.

"Well, let me balance all this despair you heard from John and Henderson by telling you that we are blessed to be here in this area isolated by the mountains. We're able to defend our families and our goods from the pillaging soldiers much better than the people closer to the cities. There hasn't been a lot of loss so far. It's true that we're not able to take our products to Gainesville like we were doing before the war, but we can mostly make whatever we need. The Home Guard has been able to drive back the Yankees who've entered our territory. We're not having parties like in the former times, but we still are able to have our church services just as before. The children are still able to attend school. The small number of slaves we had in the county have mostly stayed here and lived their lives as before."

"Yes, Papa, we are very fortunate," Jasper said. "I've seen some sad conditions in the places I've been. Houses stripped of everything that had any value, many of them burned to the ground. All of the livestock either killed or stolen, and the family members who escaped with their lives having to beg for food and clothing."

Mancil, Fate and Robert came in from school as Jasper was telling about the awfulness of war, and they stood staring at him, the horror of his words accentuated by his frail appearance. Seven-year-old Robert's eyes were wide with alarm, and Clark crossed the room to place his hand on his shoulder.

"Son, Jasper is telling about some places far from here. God willing, those things will not happen in our community. How was school today?"

Robert tore his eyes away from Jasper and looked into his father's face. His expression changed and a small smile appeared on his lips as he looked at Clark's peaceful countenance.

"I had a good time at school," he said. "Teacher let me tell the students how I can make a rabbit trap."

"Papa, he told them how *I* make a rabbit trap," Mancil said. "He can't make a trap by himself."

"Well, if he can't do it right now, he will be able to do it someday if he's learning how," Clark said. "Let's go out to the shop and build a fire in the heater, then we'll work on building the flying machine. What do you think about that?"

Robert and Mancil excitedly headed for the door. They loved watching their father work on the flying machine, and he always had scraps of materials that he let them use for building whatever they wanted for themselves.

The shop was cold and they started a fire in the heater with kindling from the box they kept filled for that purpose. Soon the fire was blazing and they added a few sticks of wood on top. Mancil opened the damper to draw air across the wood, pulling the flames over the bigger wood to make it catch fire quickly.

Clark was holding the model of the flying machine in his hands turning it one way then another as he studied the curve of the wings. After a while he held it up for the boys to see.

"Tell me what you think. Do the new wings look like an eagle's wings?" he asked the boys.

They stepped over and carefully examined the wings.

"I don't know what a wing without feathers would look like," Robert said.

"The eagles are so high in the air when they have their wings spread out like that, it's hard to tell exactly how they look," Mancil said.

Clark went over to a shelf on the shop wall and took down three maple seeds.

"Compare the shape of these whirligigs with the wings on the flying machine," he said. "Notice the small differences in the way the seed wings curve. A wing will have a stronger lift if it has a more flattened indentation on the underside. That will cause the air flowing over the wing to fall faster than the air flowing under it. That will cause it to rise."

"Papa, how do you know stuff like that?" Mancil asked.

"I love studying how things work. I wasn't much older than you when I watched the birds flying and decided that a machine could be built for a man to fly like a bird."

"Well, I know there are kites that are made to fly really high. I guess a machine could do it, too."

Cindy opened the door of the shop. "Mama said to come and get ready for supper," she said.

Mancil and Robert ran out the door and down the hill to the house.

Seeing the flying machine, Cindy walked over and looked at it carefully. "When are you going to fly it, Papa?" she asked.

"I have it about ready to test," Clark told her. "How do you like it?"

"It looks like a canoe with wings."

"It will work a lot like a canoe, but it will move through the air instead of water. The wings will be used sort of like paddles for steering a boat and to help lift it, too."

"We had better head to the house. Mama might need me to help her finish getting the food on the table," Cindy said.

"You're a fine girl to help your Mama like you do. Let me close down this heater and I'll be along in a few minutes."

Clark replaced the maple seeds on the shelf, placed his model on the back of the table, and shut the damper on the heater. He stood looking pensively across the room.

I've spent so much time working on this mission but, if it flies when I launch it, it will be worth all the effort, he thought, as he closed the door to the shop and headed to the house to join the family. It was really good to have them all together again since Jasper had come home from the Army.

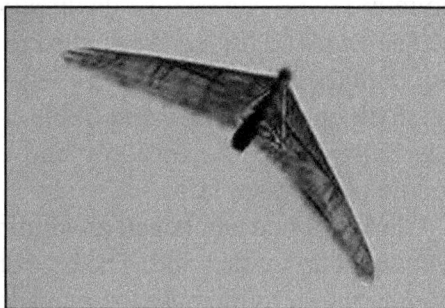

26. Catching the Wind

C lark went to his shop immediately after lunch to get his flying machine ready to take to the meadow where his cousin Jim Lance and he were to meet and see how the model would perform with its new wings. Clark's plan was to head the machine into the wind and let it fly solely from the pickup it could get without any power source. He planned to find a power source for it after ensuring that the curvature of the wings would provide good lift. He tied a cord to the model to retain control of it after it became airborne.

He stepped out of his shop and looked at the shirts hanging on the clothesline. He observed a fairly steady wind blowing from the northwest, which he thought looked adequate for lifting the model. The wings mounted on each side of the hull were covered with the corn shuck fabric he had made and were shaped with a curve that was patterned after the maple seed. The hull was made of cured river canes and looked much like a miniature canoe.

Jim came riding up in his usual exuberant way. Smiling broadly, he slid from his saddle almost before the horse came to a stop.

"Howdy, Clark. I think you picked a good day for flying your model," he said.

"The conditions look good now," Clark said. "Let's head over to the meadow and see what the model will do. This tricky mountain wind can change its direction in the blink of an eye."

They walked briskly down the hill and climbed over the three-foot high rock wall that served as a fence to keep the free-ranging livestock out of the yard and fields. They walked along the creek side halfway up the meadow and turned to cut across to the middle of the open field.

Clark stopped and began unwinding the string to let the model get caught up in the wind. The breeze had subsided and he began running into it to increase the intensity of its force. Jim stood and watched, confident that his cousin would get it up.

Shortly, the wind started blowing strongly again. Clark stopped and began feeding the line to the model, letting the wind catch the wings and lift them steadily. At once, the little craft rose and pulled the string to its end. Clark was well pleased with the height of the flight, but he quickly observed that a real flying machine would have to be built with the capability to turn its position relative to the wind's direction in order to control the path of flight.

He began slowly rewinding the string to bring the model down. Jim came across the field and held out his hand.

"Well, Clark, it looks like your new wings work really well."

"They did a good job of catching the wind, didn't they?" he said as he shook Jim's hand. "I could wish for less bouncing around, but I can't say I was surprised at that. If you watch the birds, you see them tilt their wings to control their direction when they're floating along in a breeze and it changes its course. I will have to devise a way to mimic what they do."

"How will you do that?" Jim asked.

"I think I can fasten cords to the wings and attach the other ends to pulleys that can be operated by hand to change the tilt of the wings."

"You keep working on it, Clark. You're going to figure out how to make that thing fly just right."

Clark appreciated Jim's blind faith in his ability to do it, and their conversations also sparked new ideas about ways he could accomplish what he wanted to do.

Clark and Jim had started back to the house when they saw Mancil, Fate and Robert coming home from school. As soon as the boys saw their father and cousin in the field, they turned off the trail and headed towards them.

"Papa, have you already tried out the flying machine?" Mancil wanted to know as soon as they were within speaking distance.

"Yes, I did and it works like it should."

"Will you fly it again for us to see?" Fate asked him.

"It looks like the wind may be too weak to lift it now," Clark told him, but seeing Fate's sad look, he decided to give it a try anyhow.

"Let's see what it will do," he said, unwinding some of the cord from the spool.

He tossed the model up and began running into the breeze. To his delight, the wind was strong enough at the higher elevation to lift the machine and start it climbing. The boys stood with their mouths open in astonishment.

"Papa, can I hold the cord?" Fate asked.

"Sure. You'll have to hold it tight," Clark said, handing him the spool and stepping over close enough to grab it back again if Fate should turn it loose.

After watching for several minutes, Robert wanted to take a turn holding the cord.

"He's not old enough to do it right," Fate said.

"Let him try it, Fate," Clark told him, carefully taking the spool from his hand and placing it in Robert's. He could feel the tension in the cord as the model flew at maximum height.

"Remember to hold on tight," he told Robert.

"The wind is pulling it hard, but I can hold it. Just watch me," Robert said.

It didn't take long for Robert to tire of holding the cord and he handed it back to Clark. His curiosity satisfied, he ran off to find something more exciting to do.

"Do you want to fly it for a while, Mancil?" Clark asked.

"No, sir. It looks like its flight is steady enough to be a little boring right now."

"Would you like to bring it down, then? Jim and I were ready to head back to the house when you and your brothers got here."

"Yes, I'd like to do that, Papa. Do you think I can do it without making the model crash?"

"Sure, you can. Take hold of the cord and I will hold the spool. Walk quickly downwind, bringing the line down as you go, hand over hand, while I come along behind you winding all the cord you pull down back onto the spool."

Mancil took hold of the cord and started down the field, pulling the line first with his right hand and then with his left and letting it fall behind him for his father to pick up. He was surprised at how powerfully the wind was pushing against the model. The line was taut and he wondered how his father knew about doing so many things, like how to make the model strong enough to hold together with so much strain on it, and how to attach the cord in the right place and manner where it wouldn't break or tear loose in a strong wind. He suddenly realized that neither he nor his brothers really appreciated their father's intellect. While quite a few men in the community called him a genius, there were others who thought he was wacky.

The wind suddenly died down, and Mancil felt the line go slack.

"Walk faster and keep pulling the line down," Clark said, as he dropped the spool on the ground and began running toward the model to be able to catch it when it descended enough for him to reach it. The strategy worked, and he seized the model before it fell to the ground.

"Good work, son," he called back to Mancil, who was relieved that his father had safely retrieved the model.

"I had to drop the spool. Go back to it and wind the cord evenly around it like I was doing. Keep an eye on the line and don't let it get tangled. I'll wait here and hold the line taut until you wind all the way to me."

Clark watched Mancil carefully wrapping the line around the spool as he came toward him. He was thrilled that the new wings had functioned so well and that his sons were there to see it and participate in the test.

Mancil handed him the spool when he reached him and made the last wrap around the spool.

"Papa, I am really proud of you," Mancil said as he looked closely at the model in his hand.

"I'm glad to hear you say that, Mancil," Jim said. He had been quietly observing from the sidelines how Clark interacted with his sons. In the past, he had occasionally wondered whether they appreciated their father's talents as much as they should, but now he was hopeful they were beginning to see him in a new light.

Clark called to Fate and Robert who were playing along the creek bank, "Hey, boys, come on to the house now."

As they walked toward the house, the conversation was lively about whether a man-size machine could be built that would fly like the model.

"Really, there's no reason it can't be done," Clark said. "The machine would need a source of power to lift that much weight into the air, but until such a source can be designed or found, a big balloon could be attached to lighten it. Once it is airborne, the means for controlling the direction of flight would be the same as is necessary for the model."

"There are a lot of hot air balloons around. I don't know how much weight they will support, but that might be an option," Jim said.

"They are likely quite expensive, and I don't have any spare money right now," Clark said.

When they arrived at the shop, Clark placed the model on the worktable, closed the shop door, and turned to Jim.

"Come in and have supper with us before you go, Jim."

"No, thank you, Clark. I'd better get on home and take care of a few chores before nightfall. It was exciting to see your model fly like it did today. I'm glad I got to see it."

With that, he mounted his feisty palomino and rode down the trail toward home.

Morena met Clark at the door with a big smile.

"Mancil said your model flew like a bird over the meadow this afternoon. Good for you!"

"Yes, it did. But you know me, now I have bigger plans for getting a man-size machine to fly like that. Actually, before I ever get around to starting a man-size machine, I've got to work out a design for controlling the direction of flight for any kind of machine. Jim was with us, you know, and he suggested using a hot air balloon to lighten a big machine since it will be too heavy to fly without some kind of power. So lots of angles will have to be worked out."

"You wouldn't be happy without something challenging to work on anyway." Morena gave him a teasing smile.

"I suppose you're right, Honey," he conceded, thankful she understood his curious ways.

He heard the boys in the living room telling Jasper about their experience flying the model in the meadow. He stood outside the door a few minutes to see how they would describe what they saw.

"It was way up in the sky," Robert said, stretching his hand up as high as possible. "And it was just staying up there like a bird."

"I could feel the wind trying to take it higher. It was pulling the spool the cord was tied to on our end," Fate exclaimed.

"I heard Papa say that he would have to figure out how to fix it where he could control the wings to make it go in whatever way he wanted it to. He talked to Jim about using a balloon to lift a machine big enough for a person to fly in," Mancil added.

"Wow!" Jasper said. "It sounds like I missed a lot by not being out there with you fellows."

Clark came into the room at that moment.

"We'll take it out again when you're strong enough to go to the meadow with us," he told Jasper.

"I hope that won't be too long, Papa. I can tell that I'm getting a little better."

"Yeah, you're going to be well before long. Why don't you use this down time to catch up on your reading on days when you feel like doing it? I still have some very interesting books that Uncle Joe gave me many years ago."

"Mama has already found them and brought them in for me." Jasper smiled and reached down beside his chair to get one to show his father. "I'm really enjoying the *Autobiography of Benjamin Franklin* and *Birds of America* by James Audubon. If

I don't get all of them read by the time I'm well, I plan to keep reading until I do. Also, Cindy has been getting me to help her teach Rena to read from Webster's *Blue Back Speller*. She can read a little of it already even though she's not yet five. We're proud of her and she's proud of herself. She's even trying to share her knowledge with Johnson, as though she could teach a two-year-old to read."

John and Henderson came through the back door into the kitchen, and Henderson called out, "Hey, Ma, we brought you some fish."

Morena went to see what they had.

"Those are nice ones. Where did you catch them?"

"We stopped at the Nottely River as we were coming from work. See, we cleaned them already. We're hoping you'll cook them for supper," Henderson said.

"I will do that. I think everybody would enjoy some fresh fish."

She took a big iron skillet from a hook on the chimney, placed it on the stove, and spooned some pork grease into it from a can on the side of the stove. While the grease heated, she dipped flour and cornmeal into a bowl, sprinkled salt into it, and quickly mixed the ingredients together to make a coating for the fish. She was proud of her cooking skills and her large family enjoyed the results. Even baby Johnson enthusiastically dug into his food at every meal.

Tonight would be special as they all rejoiced with Clark over the good results with his flying machine model today.

27. War, War, War

Six months had passed so swiftly, Clark thought, as he walked to the schoolhouse on a cool November morning to meet with Choestoe Home Guard Captain Marcus Thomas who had called the meeting to brief their unit on the status of the war and get reports from them about conditions on the home front. Jasper had largely recovered from his illness and wanted to go with his father to hear Captain Thomas speak. Clark was reluctant to take him, but since he had not been out in the community very much since coming home, Clark felt it would lift his spirits to be among the militia again.

When they arrived, the Guardsmen greeted them warmly, and everyone crowded into the little schoolhouse to get out of the chilly wind and await the Captain's arrival.

It was not long before they heard two horses coming up the narrow trail, and the Captain appeared with his aide, who had an attaché case strapped to the front of his saddle. They dismounted and the aide led the horses to the hitching post at the side of the building.

"Good morning, Captain Thomas. I hope you had a good ride coming over here this morning," Lieutenant Townsend said, greeting him with a sharp salute and firm handshake.

"Yes, yes," the Captain replied, returning the salute. "We spent the night at Miss Lena's boarding house, so we didn't have very far to travel."

"Come on in. I believe all the men are here. They have been awaiting your arrival, Sir."

The men stood and saluted the Captain as he entered. His aide came behind him, and they went to the front of the room and stood at the schoolteacher's desk. The Captain solemnly looked over the group while his aide removed some papers from the leather case and arranged them in front of him.

"Men," Captain Thomas quietly began, looking from one face to the other. "As you know, we have been in the fight against the Union to retain the sovereignty of our Confederate states for almost three years now. Many of our comrades have lost their lives, and some have lost their will to continue the fight and have deserted their posts. I appreciate all that you Guardsmen here in Choestoe are doing to keep your communities safe and functional as the conflict continues. Thank you for your help in capturing the deserters, as well as the raiders who so heartlessly steal and destroy the property of the absent soldiers and who cruelly terrorize widows and orphans left behind. Your service in bringing these criminals to justice is vital to the success of the Confederacy.

"I would also like to share news with you from the frontlines. Union forces have succeeded in taking control of the Mississippi River and have blockaded it totally. This is causing a hardship in getting supplies to our men in many of our military posts and in importing goods needed here at home that we can't produce ourselves. The most recent battles at Chattanooga have been fierce, and we have suffered a lot of casualties, but so have the Union forces. General Lee has managed to drive back the Yankees in many encounters, but we need to be praying for the Lord's help. If Grant takes Chattanooga, it will open the way for him to push into Georgia.

"I want to encourage all of you to hold on and keep your chin up. Don't grow weary or give up hope. We are fighting for a just cause and, with God's help, we will be victorious in the end." He paused a moment. "And now, I'd like to hear from you about how things are going here at home and try to answer any questions you may have of me."

Lieutenant Townsend rose.

"Thank you, Captain Thomas. It is very reassuring to hear from you what the conditions are like on the battlefield, although we would wish for news that an end of the fighting was in sight.

"Here at home we have captured five deserters and turned them over to Confederate authorities for trial and imprisonment. Three of them were caught after nighttime attacks upon homes where they took food, quilts and clothing. We were able to recover and return some of the goods to the rightful owners, which was a good thing since the victims were among the poorest of our neighbors. We are keeping our widows and orphans supplied with food and wood for their cooking and heating, and we're helping them plant and harvest their crops.

"Despite the blockades, we have been able to get some of the imported goods we need through private persons who can by various devious means get around the barricaded rivers and ports. They bring in supplies and peddle them undercover. But prices have gone sky-high for imported merchandise, and we can't get out many of our own products to sell and raise money for buying anything.

"While we certainly are experiencing some hardships from the war effects, we are well aware that ours are minor compared to what some folks elsewhere are having to endure. As to questions we have, I think ours are the same as for folks all over the country, ones that no one can answer: When will this fighting end? Will life ever get back to normal?"

Captain Thomas nodded in agreement. "Yes, those are questions all of us would love to have answered. No one knows when the war will end, but we do know it has changed our lives forever because we have lost family members and friends. I don't think that means we can't find happiness or build satisfying lives when the war is over. We can and will do so. Life will get back to a new normal when this conflict is finally ended. It won't be like it was in the old days, but life will be agreeable and satisfactory once again. Remind yourself of this when you're feeling discouraged.

There was silence for a few moments in the room as the men weighed what the Captain had said. When there were no further questions asked of him, he gathered the papers on the desk and put them in his bag.

"I want to tell you again that your service is important to your community and to the Confederacy," he said, bringing his speech to a close. "Continue in the good work you're doing, and God bless all of you."

He stepped over to Lieutenant Townsend and shook his hand, then continued around the room shaking the hand of each of the Guardsmen. When he had finished, he stood at attention and gave a final salute before walking out the door.

The room was immediately abuzz with the men trying to figure out if the Captain's remarks were intended to be more than just a routine update for them. Certainly, the news that Union forces were fighting to seize Chattanooga was disturbing. The battle had not been this close to their homes before.

"If the Union forces capture Chattanooga, they'll for sure head to Atlanta next," Joe Wilson said. "Atlanta has the best rail network in the South, and it is the primary means for transporting supplies now that the seaports and the Mississippi River harbors are blockaded. I don't doubt that the Yanks will want to destroy the railroads because they're our last way for receiving and sending goods."

"It sounds like we may be headed for some leaner days. I guess we'll have to learn to manufacture some things we've been able to buy ready made in the past," Wayne Brown said.

"I don't know how we'll do that," Ned Souther said. "It takes a lot of money and materials to start up a new business and I don't know anyone who has the means for doing that."

"Men, let's don't borrow trouble from tomorrow. We will need to handle situations as they arise the best we can. I think Captain Thomas gave us some good advice when he said we should remind ourselves that this war will be over eventually and things are bound to get back to normal again afterward." Lieutenant Townsend was attempting to lift the discouraging atmosphere that had settled over the room.

"I agree with that," Jesse Watkins said quickly. "Let's get the womenfolk to put on a little shindig where we can give our minds some rest from all this war talk."

Jesse's suggestion was eagerly accepted by the group as they left the schoolhouse and began heading for their homes.

"I am very concerned about the state of the war, but there isn't anything we can do to bring it to an end," Clark told Jasper as they rode home. "We'll just have to do our part here at home and keep praying for peace to be reached between the two sides."

When Clark and Jasper reached home and told Morena about the suggestion for a party of some kind to help take everyone's mind off the war, she quickly agreed.

"With Thanksgiving approaching, we have a theme for the party already available for us. The women and girls will be so excited when I tell them."

"I have to confess that I'm pretty excited about it, too," Jasper admitted. "I'm ready to see a crowd of people laughing and enjoying themselves again."

He was envisioning how pleasant it would be to spend some time with Emma at a party. His coughing from chronic bronchitis had subsided, and with an increased appetite he was regaining his weight and strength. He wanted to officially propose to her, but not before he felt confident of his ability to live a normal life and provide a means of support for her and a family. He was grateful that she had been willing to wait for him through his Army service and illness. He suspected that knowing she was waiting had been a significant aid in his recuperation. He had seen many of his fellow infantrymen succumb to depression when they got sick and didn't have anyone or anything to give them hope.

It didn't take long for news about the party to travel throughout the neighborhood. Plans were hurriedly made to hold the event in mid-afternoon in the churchyard two weeks hence.

Clark and Morena arrived at the church in a carriage with their five youngest children about two o'clock on the appointed day. Early that morning John, Henderson and Fate had ridden their horses over to help with roasting pork and turkeys to be served to the crowd. Jasper had taken a buggy to pick up Emma.

The neighborhood women were arriving with vegetables, breads and desserts to place on tables the men had set up on the west side of the building where they would be sheltered from the wind. The food they brought was very simple because the blockades had halted the supply of many of the ingredients they would normally use for cooking their special recipes. They wore dresses made mostly from homespun fabric, but they had knitted and crocheted scarves and shawls from yarn dyed in vibrant colors from the wool shorn from their sheep. It was a pleasant scene despite the hardships they were currently going through. It was apparent that the crowd was in a party mood.

As the Sullivan family arrived with their pretty teenage daughters, John immediately caught sight of Elizabeth in a long green skirt and white lacy blouse. She alighted from the carriage and pulled a yellow shawl over her shoulders, which accented her auburn hair in the sunlight. He stood beside the fire where the pork was cooking, taking in the lovely view.

"You better keep your eye on the cooking instead of ogling the girls," Henderson reminded him.

"You might want to look over there yourself, brother. Little Adeline has her eye on you and she is looking pretty good in that blue dress today."

Henderson turned to look and he had to agree with John. Adeline looked quite becoming and grown-up as she walked across the churchyard with her long skirt swinging. She and Elizabeth were headed toward the tables with bowls of food for the meal.

John walked over to them as they passed near the fire.

"How are you pretty ladies today?" he asked. "I know you have some good food in those bowls. We've been over here cooking for hours, and I'm hungry as a bear."

Elizabeth smiled at him. "You will find out soon enough," she said. "The pork and turkeys you're cooking smell good, too."

"Maybe it will be tasty, that is, if my brother over there hasn't ruined it," he said laughingly to tease Henderson.

"Wait just a minute," Henderson retorted. "This is going to be the best roast ever made in Choestoe."

Adeline gave him a shy smile. She was obviously impressed with his quick response to his brother, but she said nothing.

After they walked on, John said, "Henderson, you should have told her that she looked pretty."

"She knew what I was thinking. I didn't have to say it."

"Well, girls like to hear sweet talk, you know."

They glanced around as they heard another carriage coming and saw that it was Jasper with Emma in his buggy and the rest of the Lance family in the carriage behind them. Jasper was smiling, his cheeks rosy from the ride in the wind. He looked the healthiest they had seen him since coming home from the Army.

Jasper alighted from the buggy and helped Emma down, and they walked over to his brothers.

"Let me be the first to tell you," Jasper said. "Emma and I are getting married!"

"Well, congratulations to you both," John said, as he and Henderson stepped over to shake his hand and give Emma a hug. "Are you getting married right away?"

"It will be a few months. We decided to wait for spring. Emma wants to have a pretty wedding outside, maybe in April. We are hoping the war will be over by then."

"We are all hoping with you on that," Henderson said.

The musicians began gathering with their instruments and started tuning them up. They soon had the melodies of the sacred songs ringing across the churchyard. The sound of the talking and laughter as it blended with the music and sunlight had the desired effect on the crowd. All thoughts of war drifted away for a few hours.

28. Passion Renewed

Clark walked slowly behind a pair of oxen pulling a plow through the rough ground that he and his sons had cleared of timber and brush so they could expand their cultivated crops. Roots and rocks frequently interrupted the plow's path, and Clark had to stop the oxen to dig out the obstructions before they could continue breaking up the unspoiled ground. It was hard work.

But his mind was far away as he followed the plow. He was pondering how he could get financing to build his flying machine. The South was gradually getting back on its feet following the end of the war, but most of the men who had been wealthy in the pre-war days were now down to only enough to support their families. The few who had managed to hold onto some money or had profited from black market trading during the war and Reconstruction period were reluctant to embark on any venture that couldn't guarantee them a profit.

The war had basically destroyed the enthusiasm and optimism for a bright future that once had been the trademark of the pioneers. Jim Lance had summed it up pretty well when he was talking to Clark recently. "Sherman's sixty-mile swath of destruction across Georgia was more than just a devastation of the land; it wrecked the spirit and willpower of many of our people."

"Well, that might be true for many folks," Clark said aloud, with no ears to hear except his own there on the hillside, "but I refuse to let it be true for me. I will find a way to build my flying machine, and I will earn enough money to buy whatever materials I need that I can't fabricate for myself."

With that, he gave a mighty yank on a large, stubborn root. It broke away with a loud crack, and he went tumbling down the hill, leaving the startled oxen looking after him. When he got back on his feet, he threw his head back and laughed uproariously.

"Lord, forgive me for thinking I can do this under my own power. I humbly ask you to give me knowledge, help and strength to finish building that machine. You are likely the only one who knows just how important it is to me."

A cloud passed briefly across the sun, and he thought it looked like a wink from God. He took it to mean that the Lord would be with him in the venture and he felt peace in his heart.

He walked up the hill, took the reins in his hands, settled the plow back in the furrow, and continued with the tedious job of tilling the rugged land. Shortly, he looked toward the trail and saw Robert, Rena and Johnson coming home from school. They turned and headed up the hill toward him as soon as they spotted him plowing.

"Glad to see you kids coming to help me," he said, bringing the oxen to a halt as they arrived. "How was school today?"

"Papa, I learned about the lion and the mouse today," Johnson said.

"What did the lion and the mouse do?" Clark asked.

"First, the lion caught the mouse and was going to eat him, but the mouse begged him not to do it. He told him if he let him go he would never forget it and would give him help someday if he needed it. The lion laughed about that and asked how a little mouse could help a big lion. But he let the mouse go anyhow.

Then one day the mouse came across the lion in the woods where some hunters had tied him up until they could bring back a wagon to haul him to the king. The mouse gnawed away the ropes that were tied around the lion and he went free. This shows us that a little friend may prove to be a great friend someday."

"Well, Johnson, that is a good lesson to learn. I'm glad you listened carefully," Clark told him, then turned to the other children. "I need all of you kids to move the rocks and roots I have plowed out. Carry them to the edge of the field, and put the rocks in a row so we can make a fence with them for keeping animals out of our new field."

The three youngsters ran to the end of the field, put their school supplies down and began picking up rocks and roots. With their energy and gusto, they were able to clear the area quickly and, upon finishing the work, they began a game of tag with each other. After playing a while, they gathered up their school material and went to the house.

Clark plowed until the sun began to go down and when he arrived home he found Morena preparing a larger than normal meal. "Is today some special occasion?" he asked.

"Yes. It's Jasper and Emma's fifth wedding anniversary. They are coming over with their two babies to eat with us," she told him.

"Good," he said. "I'm going up to the shop and do some work while you finish putting supper on the table. I decided today that I'm not going to stop working on my flying machine. Maybe having to take a break from it has caused me to have fresh ideas to put into operation. I just feel a new assurance that I can work through any problems I encounter in getting it finished. I'll tell you later what happened to me out there in the field today."

Morena stood looking at him with a little smile on her face. She was delighted to hear that he was going to resume his work on the flying machine. It was more than just an interest for him she realized—it was a passion he had held for nearly his whole lifetime.

He saw from the look in her eyes that she understood his feelings and that she was willing to stand with him to the finish. He gave her a long, warm and appreciative hug before walking out the door.

When Clark reached the shop and unlatched the door, he stepped into the dust-covered room. He slowly walked over to his model, picked it up, and studied it carefully. His mind was running through the numerous things he needed to do to improve some of its functions. Using the plan of this model to make a machine large enough to carry a person should not be an insurmountable problem, he thought. Somehow he should be able to find or devise a source of power that would provide the thrust needed to propel the apparatus through the air. Now that he had the wings designed to get substantial lift from the passage of wind over them, he shouldn't need a lot of power to turn a screw propeller and generate enough airflow to lift the craft if he decreased its weight by some means. An idea came to mind: why not attach a hot air balloon to lighten the load?

He had already figured out how to add a rudder and stabilizer at the rear for steering and steadying the machine. He could also add foot pedals to operate pulleys attached to the hinged wings for providing lift and drag as necessary. He didn't consider any of these modifications to be outside his mechanical abilities to accomplish.

The unknown factor was whether he could get the necessary funds to purchase some of the materials he didn't have readily available to him. That was a problem he would have to face, but he would postpone dealing with it until he had done all he could with what he had.

He smiled as he put the model back on the table and walked out of the shop. If he could work in a little time tomorrow, he would experiment with some of his ideas. It felt wonderful to be back in action in his pursuit of flying.

29. Applying for a Patent

The attorney in Blairsville looked at the drawings and descriptions Clark placed on the table behind his desk. He squinted his eyes as he turned back to Clark, clearly wondering how the plain, humble man standing before him could have done this work. He didn't think it likely Clark would have had any schooling beyond the seven grades taught at the one-room school in Choestoe, but the documents looked like something that had been prepared by an engineering college graduate.

"Do you have a working model to be submitted to the Patent and Copyright Office with your drawings and narrative descriptions?" he asked.

"Yes, I do," Clark said.

"Did you work in partnership with anyone else to prepare these papers?"

"No, sir. Why do you ask?"

"Because these are perhaps the clearest and most professional-looking documents that have been brought to me by anyone filing an application for a patent. I am anxious to see what your model looks like. Do you have a full-size flying machine built in accordance with these plans?"

"I have one that is a lot like the model, but I don't have a lasting power source for it yet. I run the model with a mainspring from a clock, and I power my full-size machine with a mainspring from an eight-day-movement clock, which makes it possible for me to keep the craft airborne for a little while after I launch it off Rattlesnake Mountain. The airbag attached above the machine lightens the weight."

"Are you telling me that you have actually flown in the machine?" the attorney asked in amazement.

"Yes, sir. Several times."

"Well, that is mind-boggling!" he said. He stood dazed for a few minutes, finally calling to his secretary in the next room.

"I need you to make an appointment for Mr. Dyer to come back next week," he told her.

Turning to Clark, he said, "Bring your model with you boxed up and ready to ship to Washington, D.C. next week. I will have your description of the flying machine and the details of how it works typed up. We will put labels on your drawings and combine everything into one document. You will need to have two people witness your signature on the application. I want to get this moving as soon as possible. I am extremely interested in finding out how the patent application examiners will react when they get a look at your machine."

He again stared intently at Clark, wondering how much more this unlikely character knew about matters most of the country was completely unaware of. He decided to at least inform him about how the patent process worked.

"I don't know if you are aware of what the patent application examiners do. It is apparent to me that you are a widely read man, so perhaps you do," he said.

"I know a little about the approval process," Clark said.

"Okay," the attorney said, "I will tell you briefly how your application will be examined. They will test to ensure that your invention is new and not merely an obvious change to any other similar article. Beyond that, it must be clearly apparent that the invention works as described and is useful in its application. The work of the patent examiners calls for judgment and intelligence of a high order, a thorough knowledge of applied sciences. They will have to satisfy themselves that your application for a patent entitled 'Improvement in Apparatus for Navigating the Air' is valid, workable and new."

"I know my invention is valid and workable," Clark said. "My wings, stabilizer and rudder designs will positively improve the ability of the pilot to control the craft's flight path. As to whether anyone else has already patented a design like this, I can't say. I read everything I can get my hands on about flying, and I haven't seen where anyone has been granted such a patent yet."

"Well, I wish you the best of luck in your venture, Mr. Dyer, and I look forward to finalizing your application next week."

They shook hands and Clark walked out into the bright May sunshine. As he mounted his horse and turned toward home, he ambled along with mixed feelings. All these years of working on the machine he had wanted to have sole ownership of his flying machine, but he was beginning to wonder if it was reasonable for him to expect to earn the entire amount of money he needed for building it since the economy was in such terrible condition.

Building a first-class model and paying the attorney to prepare and send his application to the patent office had exhausted the meager funds he had been able to put aside, but if he should ever find someone to finance the undertaking, he would have to get a patent in order to protect his rights. He felt confident that no one had already patented a navigational design like his. Otherwise, the newspapers would be publishing stories about it. Hardly a week went by without there being a report about someone in the world attempting to fly one of the various types of aircraft, but none of them had been able to stay airborne on a consistent or continuing basis.

If I were wealthy enough to buy the parts and mechanisms I need, he thought, I could complete my full-size machine and put the finishing touches on it so it would fly like I want it to. Just the thought of flying a flawless aircraft above the earth caused a smile to cross his face. He hadn't come this far by falling into the mire of despondency and he wasn't going to let it cause him to slip down there now. He squared his shoulders and set his mind to fathoming how he might be able to accomplish the work that still needed to be done. I *will* find a way, he said to himself.

The week passed quickly with another wedding taking place in the family. This time it was 21-year-old Cindy marrying John Paul Smith. They would be living in Lumpkin County and the family would certainly miss her, especially her mother, Morena. Cindy had always been like a second mother to her four younger siblings—Mancil, Robert, Rena and Johnson—who were now teenagers and the only ones still left at home. John, Henderson and Fate married a few years after Jasper did, and they all lived close enough to visit often; but it would be a long trip for Cindy to come from Dahlonega to see them.

Clark went to his shop to begin packing his model to take to the attorney's office when Johnson came in to see him. He started to ask a different question but, upon seeing Clark boxing the model, he exclaimed, "Papa, what are you going to do with the flying machine!"

"I am going to send it to the Patent Office in Washington with an application for a patent on it."

"Will they give it back to you?"

"No, son. They will keep it."

"But how will you know how to build your big machine if you don't have your model?"

"I am keeping a copy of the drawings and descriptions that I am sending to them. And after all these years of working on the machine in my spare time, I will remember how it is made without having to refer to my papers very much anyhow."

"But I will miss seeing it here in the shop, won't you?"

"I sure will, son. But I expect to build another one with some improvements that will give a pilot greater control over how the machine operates in the air. What I really need is some way to power the propeller. I saw the big steamships on the Ohio River when I made a trip to Kentucky back when I wasn't much older than you, and they were using steam to turn the big paddles for pushing the ships upstream, against the flow of the river. I was amazed! In fact, I still have the journal I kept with the descriptions and drawings I made of the paddles while I was there. They were very similar to the blades on the water wheel at the gristmill. Of course, now the wheels of the trains are being turned with steam power, too."

"Papa, I think you're really smart to figure out all of this stuff, but some of the kids at school say their family thinks you are crazy to put so much work into building a flying machine."

"Don't worry about that, son. They don't understand how important it will be for people to be able to go places through the air without being hindered by bad roads or not having a ferry at a point where they want to cross a river or lake. They will be able to take a straight path instead of winding all around the mountains and fields. It's going to happen, and it won't be long. Men all over the world are trying to build a machine that can be controlled while it is flying through the air."

Clark paused. "What did you want to ask me before we got off your topic and onto the subject of the flying machine?"

"I don't remember," Johnson said, laughing. "I'd rather hear you talk about the flying machine anyhow."

"You're young enough that you will certainly see all of this come to pass. But as hard as it has become to get money for anything, I don't know whether I will still be around to see it happen. I hope I will."

The next morning Clark loaded the box containing the model onto his wagon and strapped it securely underneath his seat. He was planning to ride by and pick up his brother Cager and brother-in-law Frank Swain at their homes since they had agreed to go with him and witness his signature on the patent application.

Morena came outside as he reached the porch and handed him a bucket.

"I packed a lunch for you," she said. "I hope you'll stop and eat at lunchtime. It's not good for you to go so long without eating anything."

"Yeah, I'll do that," he said, smiling at her. "Thanks for looking after me. Did you happen to put in enough food for Cager and Frank, too?"

"Yes, there's enough for all three of you."

"I plan to be back by suppertime. I told Henderson that I would be gone today and he will be listening out in case you and the kids need anything. Make the kids behave themselves."

He slapped the reins lightly and the horse started down the trail. He kept his feet planted firmly against the model to keep it from sliding around as the wagon traveled over the rough trail.

After picking up Cager and Frank, he finished the trip to Blairsville with only one other brief stop to water and graze the horse, during which time the three of them ate the lunch Morena had prepared.

Upon arriving at the attorney's office, he dismounted and unstrapped the box to take inside.

"Come on, fellows," he said to Cager and Frank. "This shouldn't take long."

"Hello, Mr. Dyer," the attorney greeted him as he entered. "We have your documents ready to be signed."

"I have brought Cager Dyer and Frank Swain with me for witnesses," Clark said.

The attorney shook hands with them and said, "You fellows can sit at this table," waving his hand toward a table beside his desk.

"But first I want to take a look at your model, Mr. Dyer, if you don't mind."

"No, sir. That's fine. Take a look at it."

The attorney opened the box and took the model out very gingerly. He looked at it wonderingly and turned it in every direction to thoroughly examine each part.

"That's very clever the way you have the straps running around the pulleys attached to the wings and down to the hand controls so you can move the wing flaps. I see how another strap goes to the rudder, which you described in your application as being used to guide the machine."

"Yes, actually it's a quite simple system, but it works," Clark said.

"You don't consider it complicated, but I'm amazed that you, or anyone for that matter, could figure out how to put all the parts together in such a way that it would fly."

The attorney placed the papers in front of Clark and pointed out where he should sign.

"I want you to sign the original and two copies, which I have already signed, as you see. The original will go to the Patent Office and you and I will each keep a copy."

Clark and the two witnesses signed the document.

"What do I owe you two gentlemen for your assistance?" the attorney asked them.

"Nothing," they both answered.

"Well, thank you so much for obliging us. Your signatures will go down in history if Mr. Dyer gets a patent for this flying machine."

They smiled but their expressions showed skepticism about their names going down in history.

"Somehow I get the impression that many people don't have much of an understanding about the importance of a flying machine," the attorney said.

"That's true, believe me. I run up against many who don't. But fortunately their opinions don't hamper the machine from flying one bit," Clark said with a smile.

He stood to leave and shook hands with the attorney.

"I will let you know when I hear from the Patent Office."

"Yes, please do. I'm pulling for you, you know."

The attorney stood looking after the three men as they climbed back into the wagon, still feeling surprise and admiration at what this ordinary-looking man had been able to create on his own.

The September sun was warm on Clark's back as he swung the scythe through the tall grass in the meadow. Summer had been rainy and the grass had flourished. The cows and horses would have plenty of hay this winter if he could get it dried and stacked before rain came again. The corn had produced bountifully, too. He hoped he would be able to fatten a lot of hogs and cattle for sale this year.

He heard a horse coming up the trail and looked to see Henderson headed his way. As he neared him, he held up a letter.

"I picked up this for you while I was at the Choestoe Store, Papa. It's from the U.S. Patent Office and I knew you would want to see it right away," he said when he reached him.

Clark felt a sudden tightening in his throat. After all the weeks of waiting, what if it wasn't approved? Henderson watched his father standing there stiffly, holding the letter and struggling with his emotions.

Finally, he reached in his pocket, pulled out his knife, and slit the side of the envelope. His hand trembled as he drew out the letter and document. Immediately, his eyes fell on the first words of the first sentence of the letter: "We are pleased to inform you that we have approved your application for a patent for Improvement in Apparatus for Navigating the Air."

"Son, they approved it! They issued a patent for my flying machine!"

"Let me see what it looks like, Papa."

Clark handed him the document that was enclosed with the letter, but he stepped close by his side so he could read it with him.

UNITED STATES PATENT OFFICE.
MICAJAH DYER, OF BLAIRSVILLE, GEORGIA.
IMPROVEMENT IN APPARATUS FOR NAVIGATING THE AIR
Specification forming part of Letters Patent No. 154,654, dated
September 1, 1874; application filed June 10, 1874.

The rest of the narrative on the three pages was the descripttion taken from Clark's application. Two other pages included copies of the drawings Clark had submitted with his application, upon each of which the patent office had written:

Inventor: Micajah Dyer per P. Mauney Atty.
Witnesses: D.G. Stuart, Leo Van Riswick

"That's the Washington attorney who reviewed and approved the patent and those are his witnesses," Clark said.

Henderson reached over and grabbed his father's hand.

"You actually did it, Papa!" he said, shaking Clark's hand excitedly. "Congratulations!"

Clark bowed his head and quietly prayed, "Lord, I want to thank you. I know this could never have happened without your help."

He replaced the letter and patent in the envelope and handed it back to Henderson.

"Take this to the house with you, and put it in the bureau drawer in my room. I want to be the one to tell your mama that I got it. She has been a true supporter of this dream of mine over the years, and she should hear the news from me about the patent being granted."

Clark picked up his scythe and returned to cutting the grass. He felt a need to feel the soothing rhythm of swinging the scythe to settle his emotions, to allow his thoughts to relax from racing here and there with what might happen from this day forward. He knew he could enjoy sharing the news with Morena better when he was more composed.

30. Search for Financing

Clark walked into the office of *The Eagle* newspaper in Gainesville, Georgia, on an unusually hot July day. It was almost a year since he had received the patent for his flying machine and he was hoping to locate someone to provide financing for building a full-size machine as he had designed it.

John Redwine looked up from the papers on his desk as Clark walked in and removed his glasses.

"Good morning, sir," he said to Clark, rising to shake his hand. "I'm John Redwine, editor and publisher of the paper."

"Good morning, Mr. Redwine. I'm Micajah Clark Dyer from Blairsville. Pleased to meet you."

"What brings you down from those beautiful mountains in this hot weather? Everyone here is complaining about the heat, and it must feel especially sweltering to you after being in the cool elevations up there."

"I wanted to meet you and tell you about a flying machine I have invented and patented. I hope it will be a story you are interested in publishing, and that your story will generate a lead to someone who wants to invest with me in the venture. I have a trial model that I can fly, but I must build a heavier machine with a power source for it to function the way it should."

"Have you tried the businessmen and bankers in Blairsville to see if they have an interest?"

"Yes. I have met privately with every man of means in Union County, but none of them is willing to invest in the development of the machine. There are a few who are impressed with the idea and think it has merit, but they don't want to put their hard-earned money at risk to do it. I understand where they are coming from. We have suffered some serious financial blows in our country, and it has made everyone skittish about letting go of their money."

"Well, tell me about the machine. Where did you get the idea? And how long have you been working on it?"

"My first idea for a flying machine occurred when I was just a boy watching the birds fly. Through the years I have read everything I could get my hands on about man's attempts to fly. There have been men smitten with the dream of accomplishing flight going all the way back to Leonardo Da Vinci in the early 1500s. I myself have been working on designing and building this machine for 30 years."

"So your machine will actually fly?"

"Oh, yes, it flies! But it can't go very high or for long distances like one would if it were stronger both in its physical structure and the power source that drives it. I had hoped to accumulate enough money to buy the parts that I can't build myself, but even the materials required for constructing the parts costs more than I can pay."

"What is there about your design that is different from what other aspiring builders are using?"

"The body of the machine in shape resembles that of an eagle, and it is propelled by different kinds of devices, wings and paddlewheels that are all operated together through mechanisms connected with the driving power. The wings are hinged on the outer edges and can be raised and lowered, as well as set at an angle to propel forward or to raise the machine in the air. The paddlewheels propel the machine in the same way a vessel is propelled on water. It has a rudder for guiding it. A balloon is used to elevate it for takeoff, after which it is to be guided and controlled at the pleasure of its occupants."

"It sounds amazing! This will be a big story across the country, even around the world. I hope someone will read it and commit himself to investing in your mission."

"So are you saying that you are willing and able to pass the story along to other newspapers?"

"You bet I will. I am very excited about it."

Clark rose from his chair wearing a big smile and shook hands with Mr. Redwine.

"You have no idea how happy it makes me to know that the information will be spread far and wide. My flying machine works and needs to be constructed like I have designed it."

"Thanks for coming in, Mr. Dyer. I look forward to seeing you again soon."

Clark's heart felt light as he walked out the door and around the side of the building to the hitching post where his horse was tied. Maybe this story was going to turn up the investor he needed so badly. Surely, someone out there would be able to see the worth of his machine and the possible profit it could produce.

He decided he would camp along the trail going back home rather than spending money for a bed at a roadside inn. The nights were cool and he needed the solitude to mull over his conversation with Mr. Redwine. The fellow seemed friendly and accommodating enough, he thought. In fact, since he seemed to have so much interest in the venture, perhaps he might be willing to make an investment himself. Or maybe he had friends with similar financial means who would join with him to provide the funds needed.

Thoughts swirled in his head as he rode along and as he made a little campfire beside the wagon trail near a stream where he and the horse could drink. He soon finished the snack he had bought before leaving town, extinguished his fire, and lay down in the dark, quiet woods to rest and dream of flying.

After four days of travel, Clark arrived home, too tired to talk much about the visit he had had with the newspaper editor.

"Did the man say whether he will publish a story about your invention," Morena wanted to know.

"Yes, he said he would publish it and pass it on to other papers across the country. He said the story might even travel to overseas papers."

"I'm surprised that you're not overjoyed about that."

"Oddly, I was very happy at first, but the longer I thought about it, the more I started to doubt whether a newspaper story would produce anything beyond a passing interest in a novel idea from the readers who see it. Obviously, it will take more than a mere casual look at what I have for someone to be convinced to make an investment. I will probably have to continue meeting with prospective backers and hope to finally come across one who will actually put up some money."

"Don't bother your mind with it any more today, Honey. Sit here in the rocking chair while I go put supper on the table."

She is right, he thought. There's nothing to be gained by worrying about it. I will feel more hopeful tomorrow.

The Clark Dyer household was again in the middle of wedding activities. This time it was Mancil getting married to Rebecca Garrard. As he was returning from her house after they had worked several hours on their wedding plans, he stopped by the Choestoe General Store and found the July 31, 1875 issue of *The Gainesville Eagle* newspaper on the counter.

"Look at this," George said, pointing to an article titled "Machine for Navigating the Air."

Mancil leaned over and looked at the paper. It began: "MR. MICAJAH DYER, of Union County, has recently obtained a patent for an apparatus for navigating the air. The machine is an ingenious one, containing principles entirely new to aeronauts, and which the patentee confidently believes have solved the knotty problem of air navigation..."

Mancil stopped reading at that point and said to George, "I need to take this paper to my papa. He will be glad to see what the article says about his flying machine. Is this copy for sale?"

"Sure," George said. "I don't know of anyone I'd rather have it than your dad. Does his machine actually fly?"

"Of course, it does. But he is wanting to build a much sturdier and better powered craft than the one he has now. He will appreciate you letting him have the paper."

He handed George the money for the paper and quickly headed out the door. He trotted his horse along briskly, anxious to see his father's reaction to the article.

Arriving home, he immediately went to the field where Clark was stacking hay and held out the newspaper.

"Papa, look what I picked up at the store!"

Clark leaned his pitchfork against the haystack and took the newspaper from Mancil. A smile spread slowly across his face as he began reading the description of his aircraft given in the article. Mr. Redwine had taken care to accurately state how the machine would work. Clark chuckled a little as he read the concluding sentences: "Mr. Dyer has been studying the subject of air navigation for thirty years and has tried various experiments during that time, all of which failed until he adopted his present plan. He took an eagle for his model and has constructed his machine to imitate his pattern as nearly as possible. Whatever may be the fate of Mr. Dyer's patent, he himself has the most unshaken faith in its success, and is ready as soon as the machine can be constructed, to board the ship and commit himself to the wind."

Well, he's right about that, Clark thought. I know I have a design that will safely propel me through the air, and I am itching to get the big one built so I can prove to the world that controlled flight is possible.

"Thank you for getting the newspaper for me, Mancil. I'm satisfied that Mr. Redwine wrote as good an account as anyone could have about my machine. Did it strike you as a good story?"

"I think it is very good, Papa."

"Did you and Rebecca get everything worked out this afternoon for the wedding next week?"

"Almost everything. It's going to be small and simple. We agree on that point. And we decided that we will live with her parents until we can get a house of our own built instead of renting from someone."

"They will probably enjoy having you there since they're getting up in years and the rest of their children are gone. A place can get pretty boring with no young people around to liven it up."

"I'm going on to the house, Papa. Do you want me to take the newspaper?"

Yes, take it with you. Now that the sun is going down and it won't be so hot out here, I hope I can get the rest of this hay stacked."

"I will finish it for you if you want me to."

"No, no. I'm already dirty and there's no need for you to get messed up just for an hour or two of work. Tell Mama I'll be in soon."

The rest of the day went quickly for Clark. As usual, while he worked, his mind continued to swirl around possible ways he might be able to improve various functions of his flying machine and how he might be able to raise money to get the additional materials he needed. He felt blessed to have an interesting and challenging project to occupy his mind.

31. Prospective Investor

*T*he *Eagle* newspaper story about Clark's flying machine had generated a few inquiries which the editor sent to Clark for reply, but today would be the first time he would actually meet one of the people who had expressed an interest in talking to him. Philip Hudspeth had made an appointment to come to Blairsville to take a look at the machine, and Clark expected him to arrive around midday. He decided he would limit the man's inspection to the machine sitting on the ground; no need to get it airborne with the inherent risk of a mischievous wind causing a rough landing which could result in a lot of damage. If Mr. Hudspeth showed a strong interest after his preliminary examination, Clark planned to invite him back at a later date to see it fly.

Two years had passed since the newspaper story came out, and Clark had redesigned the aircraft both as to the type of materials used and the way the wings and rudder worked. This version was the lightest one he had built. He achieved the weight reduction by making the wings, body and rudder from bark he stripped from birch trees. Though the material was very lightweight, it was strong and flexible. He added a stabilizer at the rear of the craft that gave him increased control in steering and steadying the machine in the air.

Clark's youngest son, Johnson, had come with him to the shop this morning.

"Papa, where did you get the thought of using birch bark for the machine?" he asked.

"Well, I got the idea from the native Indians who lived here in Choestoe when I was a boy. Many of them prized birch trees for their bark because it is lightweight and flexible, and because it's easy to strip from fallen trees. They often used it to make canoes, bowls, and wigwams that were strong and waterproof but still lightweight. That is the kind of material a person needs for building a flying machine."

"How heavy is your machine?"

"About a hundred fifty pounds when it's not occupied. As you know, I am able to push it myself from the shop to the foot of Rattlesnake Mountain, but I have to use my oxen to tow it to the top of the mountain where I've cleared the take-off path and laid chestnut rails for it to zoom down and get enough speed to lift into the air. The first thing I am planning to buy if I can find an investor is an internal combustion engine to power the craft. I'm asking around now for someone who knows where the engines are sold so I can find out how much they cost."

"What in the world is a combustion engine?"

"The purpose of internal combustion engines is the production of mechanical power from the chemical energy contained in fuel, which today is usually gasoline. These kinds of engines have been used in European countries for more than twenty years, but the first one invented in America was just four years ago by George Brayton. Now, two fellows named Nikolaus Otto and Eugen Langen have teamed up and invented a more successful model which has an engine cycle with four piston strokes: an intake stroke, then a compression stroke before ignition, an expansion or power stroke where force is delivered to the crankshaft, and finally an exhaust stroke."

"Papa, I know you understand how all of that works, but it doesn't make a scrap of sense to me."

"Don't worry. If I can manage to buy one I'll show you how it works. You'll be able to understand it. It's not very complicated. It's amazing that no one thought of how to make one sooner."

Clark pushed against the flying machine and rolled it out the door of the shop. He brought it to a halt under the shade of a big yellow poplar tree. He brought out the wings and mounted them to the top of the machine. Then he sat down on a split log bench to await Hudspeth's arrival. Johnson had heard all he wanted to hear today about flying machine materials and engines.

"Papa, I'm going hunting with my bow and arrow if you don't need me to help you with anything around here."

"Go ahead, son. I expect Mr. Hudspeth to be here soon."

Clark watched Johnson go down the hill, stopping several times to pick up a stone and throw it toward some target. He was an expert marksman, and Clark wondered, as he often did, how it was that none of his seven sons had taken a keen interest in learning aeronautic principles and how to construct a craft that could be navigated in the air. Maybe the more reasonable question would be for him to ask himself why it mattered so much to him to build such a machine. He couldn't rightly answer that. He only knew that he had an unquenchable fire in his soul to build one.

The gentle breeze and soft rustle of leaves had nearly lulled him to sleep when the dogs began barking. He aroused himself to listen, and in the distance he could faintly hear the creak of wagon wheels and the thump of a horse's hoofs. He rose, stretched himself and walked to the spring branch that ran beside the shop. He bent down and with cupped hands dipped water to drink three or four times. He wiped his mouth with the back of his hand and ran his fingers through his hair. He walked over to the trail's end and watched as the stranger rode up the hill toward him, wondering what his reaction was going to be when he looked at the flying machine.

"Hello," he greeted the stern-looking driver. "I'm Clark Dyer."

"Glad to meet you, Mr. Dyer. I'm Philip Hudspeth." He stretched his hand toward Clark, who reached forward and shook hands with him.

"I assume that's the flying machine over there," Hudspeth said, pointing toward where Clark had parked it.

"Yes, sir. That's it. Let's step over and let you take a good look at it," Clark said.

Hudspeth walked over to the machine and grasped a wing and pulled the flaps up and down. His eye followed the path of the cord attached to the flap at the rear of the wing downward toward the bar to which the other end was attached. He reached into the hull and pushed the bar down, watching as the wing flap bent downward. He pulled the bar up and the flap went back up. He looked at Clark and smiled.

"That's simple enough," he said. "I can see how you could use that for getting lift or drag, as needed, and also for banking the machine to change the direction of flight. The description I was given said there was a balloon attached for lightening the air ship."

"That's right. The balloon is inside the shop over there. Of course, I have to inflate it with hot air and attach it before flying. Come over to the shop and take a look at it."

They stepped into the shop and Hudspeth saw the eighteen-foot envelope stretching almost to the rear of the building. He scrutinized it for several minutes without saying a word, then looked at Clark with an astonished expression.

"Tell me how you built this thing," he said.

"For the balloon casing, I used strips of birch bark that I softened by soaking in water so I could shape them in the curve I needed. Then after I joined them together for the frame, I covered it with taffeta fabric that I varnished with alum to make the material fire resistant. You know, it's a danger to have fire so close to the cloth when you're filling the balloon with hot air."

"I don't suppose it would be possible for me to see you fly the machine today, would it?"

"No, sir. Filling the envelope with hot air, attaching it to the machine and pulling the machine up the mountain to my launching strip would take several hours. You can let me know how interested you are in investing in the venture, and if you think it is something to your liking, we can set a date for you to be here earlier on the day of our next appointment so I can have it ready to fly when you get here."

"Do you know how much it will cost to build a machine like you envision?"

"I am estimating it will cost about six or seven thousand dollars. It will have to be larger than this one and must be made of stronger materials. It will require an internal combustion engine for the power source. I will itemize the materials needed to build it and give you the list."

"That is a lot of money to put into an investment of this kind, but I perceive that you are the type of builder who will search for the best price for good quality materials to build it with. Do you have any idea of the market for the flying machine?"

"I believe we would have a lot of men interested in buying it as soon as they see it. Men all over the world are seeking to construct a navigable flying machine. If we can be the first and best in building one, we shouldn't have any problem finding a buyer."

"You sound a lot more confident about that than I do at this point," Hudspeth said. "I need to investigate the market and see how it looks to me. I am seriously impressed with the machine you have built. Now I need to determine whether I feel we could find a buyer at a price that would give us a significant profit on the investment."

"I think when I lay out the cost of all the materials needed and then factor in a fair wage for the time I will have to spend in constructing the machine, we should be able to mark up the price to a level where we can profit very well on the sale," Clark said. "As to how hard it will be to find a buyer, we will definitely have to broadcast far and wide that we have a working air ship available for sale. We will have to target places where there are people who have an interest in flying and who can foresee the value of being first to own a machine that not only can be flown through the air but that can also be controlled as to the desired direction of flight. Of course, we will have to market it to people who have substantial financial means."

"It sounds like you not only need an investor for supplying cash but someone who is a skillful marketer, too," Hudspeth said.

"That's right. I would like to devote my time to building the machine and leave most of the sales efforts to someone else. It wouldn't be absolutely necessary that an investor be the one to do the marketing, but it will take someone with an understanding of the unique features of the machine to convince a prospective buyer of its worth."

"I agree with you. I have not considered the need for such a marketer until now, but I agree with you on the points you make."

"I think we can bring someone up to speed pretty quickly on the basics of how all the parts work and how the machine is superior to anything else available in the marketplace," Clark said. "But the challenge may be finding a salesperson with a natural ability to win a prospect over to our belief in the future usefulness of it."

Hudspeth laughed. "You are the perfect one for doing that. Your knowledge and enthusiasm about the machine is positively contagious. I can't imagine anyone hearing you talk about the machine without catching the vision of what it is worth because of what it can do."

"I would be willing to join the discussions after a prospective buyer is found, but I will need to devote my time to building the machine rather than searching for buyers," Clark said.

"Yes," Hudspeth agreed. "You are the only one who has the skills for doing that."

"If you are interested in investing in the machine and don't have anything scheduled for Thursday of next week," Clark said, "we can go ahead and set ten o'clock that day for you to come back and see me fly."

"Well, I am definitely interested in investing in it, but I will have to look over my existing obligations and see if I can free up that much money. I might have to find another person to join me in the venture. I think it will be best for me to contact you after I see how I can work out the money requirement before scheduling our next meeting date."

"That's fine," Clark said. "But I need to have a decision from you without much delay. If you are not going to invest, I will need to continue searching for someone who will."

"I understand," Hudspeth said. "It certainly was a pleasure to meet you and see your fine invention. Whatever my participation in the project turns out to be, I wish you the very best in raising the needed funds and getting the machine built."

They shook hands and Hudspeth climbed into his wagon and rode away.

Clark removed the wings from the machine and carried them back into the shop. He pushed the machine through the door and placed it against the worktable. He closed the shop door, went back to the bench, and sat down to consider what the prospects were of Hudspeth making the investment.

Clark was confident that the man appreciated the design of the machine, and he seemed to have a basic understanding of how it worked. But he didn't seem to have considered in advance of his visit what the cost of the building materials necessary to assemble the machine would be. Clark thought it would have been prudent for him to make some investigation into those factors first, as well as determining how much money he could afford to put into the project and what he was willing to invest.

Clark was uncertain whether the man's interest was great enough to drive him to either raise the money himself or find someone to participate as a partner with him. Time would tell.

How nice it would be if this financing problem could be settled, leaving him free to work on perfecting the machine without hindrance, he thought, as he headed down the hill to his house.

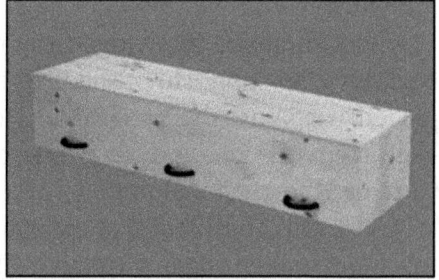

32. Another Setback

A simple pine box holding Johnson's lifeless body rested on a table in front of the pastor's pulpit at Pine Top Baptist Church. Johnson had been cutting timber two days earlier when a tree fell on him, crushing his chest and killing him instantly. His twenty-three-year-old pregnant widow, Mary, was seated between her parents, Henry and Martha Hunter, who propped her up as she alternately wept and passed out in a faint. Clark and Morena sat beside them holding their grandson, two-year-old Curtis, Johnson and Mary's son.

The remaining members of the large Dyer and Hunter families almost filled the rough, handmade pews in the little log church that men in the neighborhood had built on the craggy mountainside a few years ago. Friends and neighbors overflowed to the outside, where the October sun shone warmly upon them, moderating the chill of the wind blowing through the pine trees and across the open grave which neighbors had hand-dug in the cemetery across the narrow trail that passed in front of the church.

The pastor looked over the congregation, wondering what he should say that would bring comfort to them in the sudden and tragic passing of this young husband, father, son, brother, cousin and friend to those assembled.

He opened his Bible to Psalm 46 and began to read: "God is our refuge and strength, a very present help in trouble. Therefore, will not we fear, though the earth be removed, and though the mountains be carried into the midst of the sea; though the waters thereof roar and be troubled, though the mountains shake with the swelling thereof. There is a river, the streams whereof shall make glad the city of God, the holy place of the tabernacles of the most High. God is in the midst of her; she shall not be moved. God shall help her, and that right early."

"Mrs. Mary, I know it is hard for you to imagine right now how you can possibly go on without your husband," the pastor said. "But we have a great God who will help you and strengthen you day by day. You have a little son at your side to help you carry on, and soon another child will join you. Your parents, and his, will also help you through your grief. Talk to the Lord often. Cry when you feel like it and tell the Lord how much you miss Johnson, but also tell your loving heavenly father that you trust him to take away your sad feelings and give you joy again. Another Psalm tell us, 'Weeping endures for a night, but joy comes in the morning.' You will have joy in your life again. Believe me, and more importantly, believe God's word.

"Brother Clark and Sister Morena, I know the heartbreak you feel right now in having to give up your youngest son. I offer you the same counsel that I just gave Mary. Talk to the Lord and tell him how it hurts to have Johnson gone. But I want to assure you that you will also find that heavenly gift of 'joy in the morning' that shall come after your days of grieving have passed.

"Brother Henry and Sister Martha, I know that you were close to your son-in-law and will miss him greatly, and I know that it will give you great pain to see your daughter grieving for him. But you can rest in the same comfort of knowing that it is for a season, which means that it will pass. Of course, none of you will ever forget him, but God will bring you comfort and peace to replace your pain.

"To all the rest of the family members, I hope you, too, will pray for God to comfort your hearts in the loss of this brother. He was a fine young Christian man and his death is a loss to the whole community. But remember what King David said when his child died: 'He cannot come back to me but I can prepare to go to him.' All of you can prepare to go to be with Johnson in

Heaven someday, too. Just think what that reunion will be like—together again with no parting to ever come again!

"I will be praying for all of you in the days ahead and ask that you let me know any time you feel the need to talk about your grief or any other need you may have in your life, not because I think I have all the answers but because I know One who does. He is a friend that sticks closer than a brother."

The pastor closed with a tender prayer and the pallbearers carried Johnson's body in the pine box out of the building to the gravesite with the mourning family and friends following. A crow cawed overhead as the pastor stepped to the open grave and spoke.

"Ecclesiastes 12:7, 'Then shall the dust return to the earth as it was: and the spirit shall return unto God who gave it.'

"Family, since it has pleased our heavenly Father, who loaned Johnson to us for this short time, to take him back to Himself, we commit his body to the ground. Looking for that blessed hope, when the Lord Himself shall descend from heaven with a shout, with the voice of the archangel, and with the trump of God; and the dead in Christ shall rise first. Then we which are alive and remain shall be caught up together with them in the clouds to meet the Lord in the air; and so shall we ever be with the Lord. Wherefore comfort one another with these words. Amen."

He shook hands with all of the family and they turned to go home.

Clark and Morena began their ride quietly as they left the cemetery.

Then Morena said, her voice breaking again and again as she spoke, "On the day Johnson was born, so tiny and frail, many people thought he wouldn't live. But I remember how determined I was to take such good care of him that he would just have to pull through. I was filled with joy when he began to grow and thrive. He brought so much joy to the whole family all of his life. His marriage to Mary made me so happy. Her temperament was just suited to his. And little Curtis! Oh, Clark, how sad that he now will have to grow up without his daddy! And poor Mary, soon to have another baby that will miss having its daddy, too. I just don't understand why he had to go!"

Morena was overcome by her sorrow and lay her head on Clark's shoulder, sobbing pitifully.

Clark stopped the horse and wrapped his arms around her. "It's okay, Honey. We will all stick together, and with the Lord's help we will manage to get through our grief and help Mary the best we can to get through hers as well. We will help her manage for herself and the babies. The Lord sees our tears and he will give us comfort. We can depend on him."

Clark's own heart was heavy as lead, but to his surprise he felt it lighten somewhat as he comforted Morena. He had almost started to tell her how much Johnson meant to him and the ways in which he would miss him. Now he was glad he hadn't done that, realizing that it would have made both of them even sadder.

"Thank you, Lord," he began praying aloud. "We look to you to give us comfort and peace once again. Show us how we can give comfort to all of our children. We know there are many things we cannot understand in this life, but we don't have to understand. We just have to trust you to lead us in the way you want us to go. Fill us with your love so that we can comfort others with the comfort you have given to us. Thank you for all the blessings you have given us throughout our lives. Help us to always be aware of our blessings and to always give you thanks for each of them. We love you, Lord. Amen."

Morena dried her eyes and kissed him. "Honey, that's just what I needed. Thanks for shoring me up."

They finished the ride home, soothed by the rhythmic thump of the horse's hooves on the dirt trail, and their thoughts centered upon how they might be able to comfort their children.

Fall turned to winter, and winter to spring. Mary moved into her parents' home and gave birth to a baby girl, whom she named Charlotte. In time, she managed to rise above her grief to some degree, and she smiled more often as the months passed. She came to visit Clark and Morena frequently because they had grown to love Curtis more deeply after Johnson died, and now Charlotte was stealing their hearts, too.

Clark was continuing his search for an investor in his flying machine. Like Philip Hudspeth, his first prospective backer, throughout the years since then a number of men had come and looked at his airplane with interest and admiration, but none had made a solid offer to fund the building of it. He was discussing his frustration over the situation with his friend, John Rich, who was a county commissioner.

"John, living in a remote area like we do, a fellow doesn't have any way to gain access to financiers who are knowledgeable about science and inventions. The reason these fellows won't put their money into the venture is because they don't understand the principles behind flying, nor do they see how this method of travel is going to transform the way business and warfare will be conducted all over the world. I can see it clearly but they can't seem to grasp it."

"I have an idea," John said. "I am a good friend of the *Athens Banner-Watchman* newspaper editor and publisher over in Athens. I will send him a letter introducing you and explaining your invention and the need for an investor. I will ask for his help in finding someone for you."

A few days later, John came over to Clark's house with his draft of a letter to the editor. He handed it to Clark to read.

> April 19, 1885
> Mr. Editor:
>   Mr. Clark Dyer, of this county, thinks that he has succeeded in making an air ship. He says that he can sail through the air, but has not means to get material of sufficient strength to complete the machine. He has had it patented, and he now only lacks capital to complete the machine, and he wants capitalists to engage in the matter. He believes if he can get scientific men to examine his machine, they will come forward and aid in its completion, and as you have always been a friend to science and inventors, he wants you to take such steps as in your judgment will bring his invention before the public generally.
>   Mr. Dyer is in earnest, and has, no doubt, made something that he can propel through the air. If you think it possible to

get men of means to investigate the matter, I would be pleased if you would do so.

Mr. Dyer has worked thirty years on his machine. He is not crazed, but is in dead earnest, and confidently believes that he has solved the problem of aerial navigation. He is not a crank nor a fanatic, but is a good, quiet citizen and a successful farmer. Anything you may think proper to say or do in his behalf, I will take as a personal favor. He does not want to humbug anybody. After examination, if he has nothing he wants nothing.

John M. Rich

"Your letter seems appropriate to me," Clark said. "I certainly hope the editor will be able to help me make a connection with someone who will understand what I have and will be interested in putting up the needed funds."

"I hope so, too. You have tried hard for a long time to get this done," John said, as he folded the letter and slid it into the envelope. "I will put it in the mail today."

"Thanks, John. You're a friend indeed to me."

Nearly two weeks later, Clark was planting corn in his lower field when John Rich rode up.

"Howdy, Clark. I got the *Athens Banner-Watchman* paper today and I wanted you to see what the editor put in it last Tuesday."

He handed the newspaper to Clark, who looked at it a few minutes, then asked, "Is that all he is going to do to help find an investor is merely print the letter you wrote him?"

"I don't know, Clark. I was surprised and disappointed when I saw it," John said.

"Well, you tried to help, but I don't believe printing the letter alone is apt to accomplish much. I was hoping he would discuss the matter privately with several prospective investors who had sufficient knowledge of the present developments in aerial navigation to assess my machine and understand its worth."

"I understand your frustration, Clark. You can't imagine how disillusioned I am in his weak efforts at fulfilling my request. I have done favors for him in the past and he should have done more to help here. Just printing my letter is certainly not what I expected of him."

They stood several minutes, shoulders slumped, heads bowed, saying nothing. Then John climbed back on his horse.

"I know it's not much consolation to you, but I will keep my ears and eyes open to the people that I come in contact with to see if any of them might have an interest in talking to you about your venture. I hope you won't hold it against me that the editor let you down."

"No, I don't hold it against you, John. Thank you again for the effort you made."

With a heavy heart, Clark went back to planting corn and trying to keep his spirits up. Why did doors keep closing in his search for funding? There must be someone out there who would love to be involved in this project. But where could they be?

33. An Unpleasant Turn

S tymied by a lack of funds to continue his work on a stronger airplane, Clark was devoting the time in his shop these days to building other things—some practical, others merely unique ideas he had about new ways of doing things. He thought back to how, when he was a young man, he had hollowed out logs and placed them end-on-end within a ditch he dug to his house from an ample spring on the hillside. He then channeled the spring water into the pipes which he connected to a barrel in the kitchen and back out through the rear wall of the house to another barrel. He ran a log pipe from the outside barrel to a small stream that ran beside the house to the main creek.

It was an ideal way to supply the water needed in the kitchen, as well as for the family's personal bathing. It also provided water for their use outside to wash clothes and clean up behind messy jobs. At Morena's request, he had also piped water through the new springhouse he built near their house where it was more convenient for her to keep items requiring cool temperatures, such as milk, butter and eggs. As an added bonus, having water available beside the house to play in was a never-ending pleasure for the grandchildren when they came to visit.

That project was the kind of thing Clark loved doing, making something from materials he had on hand that no one else had thought of making or for which they lacked the necessary skills to create.

One of the unique projects he was presently working on was building a perpetual motion machine—something that would move under its own power without outside stimulus. If he could perfect the machine, a larger version of it could also be built to provide free power for operating other machinery or objects.

He decided to begin his experiment by attaching rods six inches long at even intervals around a sixteen-inch-diameter wooden wheel. He was carving grooves in each rod deeply enough to hold lead balls. He planned to leave the ends of each rod closed to stop the lead balls when they rolled down the groove to the end of the rod as the wheel turned. He reasoned that with the weight falling to the bottom of the rods when one side of the wheel turned over, the balls in the rods on top would slide to the center of the wheel. Therefore, he believed, the alternating position of the weights would cause the wheel to keep turning perpetually. It seemed to him that if he balanced the wheel properly, it should work.

The daily demands of his life left little time for working on the experiment, but while he was occupied with mundane work, he often turned ideas over in his mind regarding possible methods to try the next time he was in the shop. Today as he sat outside under the poplar tree carving grooves in the rods, he saw his grandson, Matthew, coming up the hill towards him.

"Howdy, young man," Clark greeted him. "Come sit on the bench with me and tell me what you've been doing."

"Hi, Grandpa," Matthew said. "What are you making?"

"I'm trying to design a machine that will keep running forever without needing water or wind to push it, no horse to pull it, and no wind-up spring to turn it."

"But how will a little stick like that do it?"

"This is only going to be used in a small model for testing whether my idea works."

"Grandpa, you are always working on things that nobody understands."

"But you see that I worked on the flying machine until I got it to sail through the air. Don't you think that proves my ideas are good ones?"

"Yes, I do. That's why I came up here to the shop to see what you are working on. I knew it would be something interesting."

"Always pay attention to how things work, Matthew. See if you can imagine another way of doing a job that would be easier or quicker. Try to have an interest and enjoyment in whatever you do. Just because some of your work is dull doesn't mean that your mind can't be having fun as you imagine all kinds of possibilities in how the same job might be accomplished by doing it a different way."

"I want to go in the shop and see how you are making the perpetual motion machine," Matthew said.

"There's not a whole lot to see yet, but come on and I will explain it to you."

But when Clark stood up, his head began to spin and he suddenly felt faint. He sat back down on the bench and dropped his head into his hands. Matthew went on into the shop without noticing that Clark wasn't following him.

Matthew walked into the shop and spotted the wooden wheel with some of the rods lying beside it. Clark had already carved the grooves into some of them. He turned to ask how the rods would be placed on the wheel when he discovered his grandfather wasn't behind him. He went back to the door and saw him bent over on the bench. He ran to him.

"Grandpa, what is wrong? Are you sick?"

Clark said weakly, "Go to the house and get a cup of water for me. Hurry! I'm feeling faint."

Matthew didn't need to be told twice. He raced down the trail to the house and into the kitchen. As he grabbed a mug from the cupboard and began filling it with water, Morena came in and saw the agitated look on his face.

"Matthew, what is wrong?"

"Grandma! Grandpa is sick! He's bent down on the bench outside the shop."

"Run quickly and give him the water. I'm coming as fast as I can get up the hill to him."

Morena hurried up the trail and was completely out of breath when she reached Clark. She sat breathlessly on the bench beside him as he drank the water that Matthew had brought him. She rubbed his back and neck. He was so pale!

"Honey, are you hurting anywhere?"

"My heart is fluttering and I feel very strange, very weak and sick at my stomach."

"Can you get to the house if Matthew and I support you on each side?"

"Let's try it. I need to lie down."

Then slowly and haltingly they made their way down the path to the house. They helped Clark inside and assisted him with getting onto the bed. Morena removed his shoes and put an extra pillow under his head.

"Matthew, run home and tell your daddy to get the doctor to come here as fast as he can. Tell your mama to go see if Granny Wilda can come, too."

Morena brought a warm damp cloth and wiped his face. She sat down beside the bed and softly hummed a hymn. The cold hand she was holding seemed to be getting a little warmer, and she breathed a prayer for the Lord to touch and heal her husband's weak and irregularly beating heart.

She thought of how Granny Wilda often made an herbal tea for giving to patients, and she rose and went into the kitchen to put another stick of wood in the stove so she could bring a kettle of water to a boil in case it was needed. She stirred the pot of chicken vegetable soup she was cooking for their supper and was glad that it had been her choice today since it was one of Clark's favorite meals.

She heard the dogs barking and went to the door and saw Granny Wilda coming. She quieted the dogs and waited for Wilda to get to the porch.

"I'm so glad you were able to come, Granny Wilda," she said. "Clark's heart is racing and beating very faintly. I hope you can give him something to help get it back in rhythm."

"I brought some dried foxglove leaves in my bag so I could make tea for him," Wilda said. "It sounds like he has heart dropsy from what Rebecca told me. So many of his family members have been inclined to have that and it is hereditary."

"I thought there might be a need to make tea, so I have a kettle of hot water on the stove now. Come on in and fix the tea and check him over. I am anxious to see what you think about his condition."

"Bless your heart, Morena," Wilda said. "I've come here to help your children many times through the years, and I've always found you to be especially useful in anticipating what I will need. I really appreciate your being ready to lend me a helping hand."

"I don't know what we would have ever done without you all those times, Wilda. You have usually been able to get here quickly, while it takes a good bit longer for someone to go and get the doctor."

They went into the bedroom and Wilda took Clark's wrist in her hand. His pulse remained very weak and his heartbeat was still racing and irregular. She felt his brow and it was cool. She felt certain that he was suffering from heart dropsy.

"Clark, I'm going to fix some tea for you, which I think will get your heart working right again. Relax as well as you can. It won't take long to make tea. Morena already has the water heated."

Wilda hurried off to the kitchen and Morena sat on the bed beside Clark, stroking his head and assuring him that he would feel better soon. She prayed quietly for him. She thought his color looked better, but he seemed so feeble. She wanted to ask him if he was feeling better, but speaking seemed to be an effort for him, so she didn't.

Wilda came back in a short while with the foxglove tea.

"Let's put another couple of pillows at his back and head so he can sit upright to drink the tea," she said.

When they got him situated, he slowly sipped the tea. After drinking half of it, he handed it back to Wilda.

"Try to drink all of it, Clark. It will make you feel better," she said.

He took the cup back in his hand and started sipping it again, but before he finished it, he handed it back to Wilda.

"No more," he said and sank back into the pillows. "Move some of these," he said, waving toward the pillows.

Morena moved two pillows away and straightened the other two to lower his head. He lay back with his eyes closed.

"Is there anything else I can do that might help him?" she asked Wilda.

"No. I think we need to just wait for the doctor now and see what his opinion is of what we should do."

They didn't have to wait long. The doctor arrived soon and hurried inside to examine Clark.

"What has he taken since his attack?" the doctor asked.

"I gave him a cup of foxglove tea and he drank about three-quarters of it about ten minutes ago," Wilda said. "We've kept him lying down and quiet for almost an hour."

"That was a good choice, I think. It appears to be heart dropsy. Is this the first attack he has had of that?" he asked Morena.

"Yes, I believe so. If he had a problem with it before, it must have been mild enough for him to pay no attention to it."

"He will need to avoid any heavy lifting or exerting himself very much for at least a couple of weeks."

The doctor turned to Wilda. "Make a quart of foxglove tea for Morena to keep in the springhouse. He should drink a half cup of it every day until he has taken all of it. I will come back in ten days to see how he is doing," he said to Morena. "Of course, you should send for me if he doesn't get better."

He shook hands with the ladies and said to Clark, "Take care of yourself, sir. I know Miz Morena will look after you real good."

As soon as the doctor left, Wilda went into the kitchen to make tea to leave with Morena. She was pleased that the doctor had agreed with her opinion about what Clark's problem was and the appropriate treatment. He had always been respectful of her when they both were involved with the caring of a patient. After reading everything she could get her hands on about diseases and medications and spending years treating nearly everyone in the community, she had become a valuable and quick help whenever a medical need arose.

Morena invited Wilda to stay and eat with them, but she declined, saying she needed to get home and put supper on the table for her husband. He was getting up in years and she almost ran the household by herself, even after meeting all of the medical needs that were brought to her for handling.

Morena waited for Clark to get about an hour of sleep, then took him a bowl of chicken vegetable soup to see if he could eat any. He surprised her by eating nearly all of it. He looked better and said he believed his heart was pumping better. She felt his pulse and decided it had slowed a little and was somewhat stronger.

After getting a reasonably good night of sleep, Clark was able to get up and eat a light breakfast with Morena the next morning. But several days passed before he regained much strength. When the doctor came back ten days later to examine him, he declared his heart to be strong and beating regularly again.

"But let me give you some advice for the future, Mr. Dyer. You should not do any more hard work. Light work will be fine, but heavy lifting or anything requiring much exertion should be avoided. You should even ward off emotional stress. Your heart will go out of rhythm again if you don't take it easy."

"That doesn't sound like the kind of life I'm used to living," Clark said.

"Of course, it's not. But you don't have the strong heart you used to have. If you want to protect your health, you will have to change your lifestyle."

After the doctor left, he sat in his rocker on the front porch a long time, thinking over this turn of events. He thought back to how his Pa had begun having this same kind of heart problem when he was about his age. He remembered how scary it was to both him and Ma when they saw any indication that he was on the verge of having another attack. From that time on, they kept an eye on him all the time.

The thought had sometimes entered his mind that this might happen to him in his old age, but he certainly didn't consider himself to be an old man yet. He still had so much he wanted to do before he left this world, mainly to get someone to help finance the cost of building the larger and better flying machine.

He lifted his heart in prayer: Please show me some way to accomplish my dream, Lord. I want so badly to get it done before my life is finished.

Morena noticed that he had been on the porch a long time and went out to see about him.

"Honey, are you okay?"

"Well, it's going to take a lot more thinking and a lot more praying to get comfortable with what the doctor told me. I need to decide who will take care of you when I'm gone and what I'm going to do with the flying machine."

"You wait just a minute, Clark Dyer!" Morena said, putting her hands on her hips and planting herself squarely in front of him.

"That doctor didn't say you were about to die—he said you have to change your lifestyle to one with no heavy work and no stressing out over things. I don't want to hear any more 'death talk' from you, and I don't want you thinking it either. You can make your plans for doing less heavy work, and you can make peace with yourself about selling your flying machine. As for what will happen to me, we don't know but what I might go before you do. If one of us is left behind for a while, the kids will take care of us, just as we did with our parents. Nobody gets to choose every step of their life, for heaven's sake!"

Clark stared at the spectacle of her standing there in that position with her dark eyes flashing and her pale face framed by her silvery hair. It brought to his mind a day long ago when they were newlyweds and she had pitched just such a fit over something he had done. Her fit gave rise at that time to an expression within the family, "having an Owenby fit." He tried to choke back his laughter as he looked at her, but it broke forth anyway, and he rocked back and forth in his chair, peals of laughter booming through the air.

Predictably, this added to Morena's fury. But as Clark's hilarity continued uncontrollably, her expression slowly cracked, and finally she collapsed into the chair beside him, joining in the laughter. Just then, Henderson came down the trail that ran in back of their house. He had come to see if his father was doing okay and coming upon this scene on their front porch, instead of being amused, he was seriously alarmed.

"What's wrong, what's wrong?" he demanded to know.

Their surprise at seeing him and his obvious concern about their welfare sobered them immediately, but they were out of breath from the laughter and could hardly speak.

"Just having a little laugh, son," Clark finally said, wiping his eyes.

"Wait a minute now! You were scheduled for a checkup by the doctor today. What did he say to you?"

"Said my heart was strong and in rhythm again."

"The doctor told him he had to avoid heavy work and stress from now on," Morena added.

"Is this your idea of taking it easy, Papa? I could hear you two laughing half-a-mile before I got here! Laughing that hard would wipe me out so badly I couldn't walk across the floor. I know it can't be good for your heart."

"Well, it might not have helped my heart, but it sure did help my spirits," Clark said. "Come on in and visit with us. How is everything going with you?"

"We're fine. I'm glad you got a good report from the doctor today. I hope you're going to pay attention to what he told you."

"Yeah. I guess I'll have to. It looks like, however, that I have reached the place where I'll have to do something with my flying machine if a new one is ever going to get built. Do you have any interest in me giving it to you? You'd better speak now. I'm going to take action on selling it right away."

"Papa, I really don't have any natural skill at that sort of thing. I believe you should sell it if you can find a buyer."

"Maybe I'll take a trip down to Gainesville and talk to the newspaper man again about it. He has always seemed to have an interest in it, and he might just buy it from me. It would really take a load off my mind if I could find someone to carry on with the project."

After Henderson went home, Clark mulled over the conversation he'd had with him. He wondered if the Lord had used it to lead him in the path he wanted him to go. Perhaps so.

34. Final Flight

E ven though Clark was very careful about involving himself in stressful activities after his first bout with heart dropsy, he nevertheless continued to have occasional attacks with heart arrhythmia. He finally had to face the reality that he couldn't continue making flights in his aircraft with his health becoming precarious. He even wrestled with the possibility that he might not have many days left to live. If there was to be some assurance that his new machine would be built, the best chance, as he saw it, was to find a buyer. He thought over all of the prospective purchasers he had encountered over the past decade and a half and decided that John Redwine in Gainesville would be the man most likely to come through with an actual purchase of the aircraft.

A few days later he began gathering up some items that he felt Mr. Redwine would be interested in seeing. He was spreading them out on the bed in the room where their children had slept in bygone years when Morena came to the door.

"What are you going to do with those things?" she asked.

"I'm going to pay a visit to Mr. Redwine and talk to him about buying my flying machine."

"When are you thinking about going?"

"If Henderson can ride down with me, I would like to go next Monday. I know here at the end of January the weather can turn bad, but there's not much any of us menfolk can do outside right now anyhow. I might as well use the time to see if I can nail down a sale of my machine. I also want to go for a flight while I'm still able to do it. You know, I'm a little weaker than I used to be. It takes a lot of energy to operate all of the pulleys and pedals when I fly."

"I know it does, and I don't think you need to be flying right now."

"Oh, don't be concerned about me doing it. I just need to feel the thrill of being in the air again. It looks like we're going to have some decent weather the rest of this week, so maybe I can plan to do it tomorrow or the next day. I'll check with Henderson and Mancil and see if one of them can come over and help me with getting the machine up the mountainside to the runway."

"I wish I could talk you out of it, but I know how hard headed you can be when you set your mind to do something," Morena said resignedly.

When Clark talked to Henderson and Mancil the next day, it turned out that both of them were available to help and also they both wanted to see him fly again. They convinced him to make a little party out of the occasion by bringing their wives and children. They would kill a few chickens to barbeque and bake some potatoes. Their wives could cook some leather britches and bake dried apple pies. They wanted all of the extended family to be notified of the plans and invited to come for the occasion.

Morena liked the idea of getting the whole family together. The short winter days could sometimes feel dreary, and this would be a good excuse to liven everyone up. She knew it would help keep her from worrying about Clark's last flight if she was surrounded by family. This little break in the cold weather was a perfect time to do it. Although it was short notice for everyone, they set the occasion for the next day.

The sun rose brightly the next morning and the children began happily playing together as soon as they arrived at their grandparents' home. The women started the food preparation, while the men were at the workshop attaching the wings to the flying machine and loading it onto the wagon so they could haul it up Rattlesnake Mountain.

The brisk breeze blowing up the valley was just the right speed for the machine to lift off as Clark reached the end of the runway.

"Tell everybody to get out on the field where they can see Grandpa take off," Matthew yelled to the children. "Tell our mamas to come, too."

They all hurried to the sunny side of the field where they would also be sheltered from the breeze while they waited for the flying machine to come swooping down the tracks.

Seated in his aircraft, Clark looked down the mountain at his large family waiting to see him fly. He prayed that it would be a safe flight, one the family could joyfully remember him by, and one that he could relish as he thought back upon it in the days ahead. He felt a flash of sadness thinking of his future without a flying machine. But being ever the optimist, he refocused his vision toward the day ahead when his design of the larger machine would be built and driven by a powerful engine. It would be able to take flight on a level runway. What a day that would be!

He looked at his strapping sons standing on either side of the machine there on the mountain runway, holding the brakeboards and awaiting his communication that he was ready for takeoff. He lifted his hand and signaled them to pull back the boards. They immediately withdrew the boards and gave the machine a shove.

It sped down Rattlesnake Mountain as Clark pulled the cords to lift the wing flaps. Before he reached the end of the runway, the wind under the wings had lifted the machine into the air. He pressed on the pedals to tilt the wings, making a turn to the right toward the upper end of the field. He shifted his weight with the turn and pulled the cords to let the wing flaps drop down to level again. The eight-day wind-up clock spring that he had attached to the propeller on the front of the machine was turning it rapidly, creating enough wind to keep the machine at an elevation of about fifty feet.

Clark leaned back and relaxed. He smiled and waved to the family below, pleased that all had gone well with the takeoff. He estimated that he could make a couple of circles over the field before the clock spring would wind down forcing him to land. He was keeping watch over the propeller to be aware of when it started to lose speed. He let the feeling of bliss flow through his whole body, drinking in the delight of sailing like a bird through the air.

His family below was awestruck. It was the best flight he had ever made. Morena breathed a prayer of thanksgiving for his safe takeoff. She knew he was having the time of his life soaring in the air over his fields, but she also profoundly felt the loss she knew he was already experiencing about the coming sale of his beloved machine. It was a price he would pay for being decades ahead of others in his understanding of the principles of flight, while at the same time being handicapped with insufficient funds to build the machine he envisioned.

When Clark saw the propeller begin to slow down, he steered the flying machine to line up with a spot in the pasture that he knew was the smoothest place for landing. Suddenly, a gust of wind hit the little craft, setting it to rocking. The grandchildren screamed, and although the grown-ups were not screaming, they were very anxiously watching as the machine was being buffeted by the wind.

Clark wasn't worried about the pounding though. He angled his wings against the attack and waited for the gust to subside. He proved to have judged the situation correctly, and in a few minutes the wind calmed. He easily steered the craft toward the landing spot, as he lifted the wing flaps and pointed the nose downward. He bumped to a landing only a foot past the place he had aimed for. Not bad, not bad at all, he thought.

The family came running to him, hugging him and congratulating him on the good job he had done. Henderson and Mancil brought the wagon over to haul the flying machine back to the shop.

"Papa, that was fantastic!" Henderson said, leaping out of the wagon and shaking his father's hand vigorously. "Weren't you worried when that gust of wind hit you?"

"No, not at all," Clark said. "It probably sounds crazy, but it made the flight even more exciting to have the challenge of helping my machine outmaneuver that rogue wind."

Henderson laughed and shook his head. "Papa, you ain't right in the head!"

The family headed back up to the house to enjoy the big meal and more fellowship.

The next week, Clark and Henderson traveled to Gainesville to talk with John Redwine about his interest in buying the flying machine. The weather had turned cold and windy. Although they were dressed warmly and kept their wool mufflers wrapped snugly around their necks and faces, they still felt the bite of the wintry breeze as they traveled in the covered wagon along the mountain trail.

"Papa, are you sure this weather isn't too severe for you?" Henderson asked Clark.

"It's a lot colder than I expected it to be, but I'm doing okay."

"Well, let's stop at the first lodging place we come to. I'd feel better if we didn't stay outside too long."

"That's a good idea, I think," Clark said.

When his father agreed so readily to an early stop, Henderson felt even greater concern about the wisdom of making this trip with him in the frigid weather. He earnestly hoped tomorrow would bring bright sunshine and calm winds.

As they descended the south side of the mountain range, they reached a little log house that took in overnight lodgers. They stopped and knocked at the door to inquire about the availability of a room. A little old man came to the door.

"We're in need of a room for the night," Henderson told him. "Do you have one we can rent?"

"Why, shore! Ain't too many folks traveling in this cold weather. Come on in."

"I don't think we would have planned a trip right now if we had known the weather would get this cold," Henderson said.

"Let me get our bags from the wagon and see where we can put up the horse for the night."

"Come with me and I'll show you to the barn. Is this your father? He can go in the front room and sit by the fire so he can warm up."

When they came back in, they found Clark almost dozing as he sat by the crackling fire.

"You feeling okay, Papa?" Henderson asked.

"I'm fine. This fire feels so good, it's making me sleepy."

"I don't guess you folks have been to supper, have you?" the man asked.

"No. We'd be happy to get supper from you included with our lodging if that's possible."

"Sure. We'd be happy to have you. I'll go tell my wife to put on a little extra for you."

It didn't take her long to get the meal on the table, and she and her husband seemed happy to have their company. She noticed that Clark seemed very sleepy as they were eating.

"I have a pot of sassafras tea brewing on the back burner of the stove," she said. "Would you like a cup of it before you turn in for the night? We sometimes drink it at night when the weather is cold."

"I'd like about a half cup, thank you," Clark said. "I sometimes drink it at night, too. It seems to help me sleep better."

Clark and Henderson didn't spend much time socializing with the old couple after supper but turned in early. Their bed had a feather mattress topped with plenty of warm quilts, and sleep came quickly for both of them.

After a generous breakfast the next morning, they donned their warm coats and scarves and continued on towards Gainesville. The wind had abated and the sun rose brightly.

"Looks like the Lord is smiling down on us this morning," Clark said. "Of course, I must add that he smiled on us last night, too, letting us find such a good place to lodge. That couple was so kind and peaceful. I slept about as good as being in my own bed."

"I'm glad to see you feeling better today, Papa. That cold wind was about to get the best of you yesterday," Henderson said.

Three days later, after having more agreeable weather for traveling, Clark and Henderson arrived at the newspaper office in Gainesville. John Redwine rose from his chair to greet them when they entered.

"It's good to see you again, Mr. Dyer," he said.

"It's good to see you, too," Clark replied. "This is my son Henderson."

"Are you following in your father's footsteps with the flying machine, Henderson?" Mr. Redwine asked.

"No, sir. I didn't get an aptitude for it."

"Actually, none of my seven sons got a taste for building machines and taking to the air," Clark said. "As a matter of fact, this brings up the reason for our visit today. I want to talk to you about your interest in buying my patent, model and flying machine. I have started to have some problems with my heart and don't feel that I can safely continue flying. I haven't been successful in locating anyone who will invest in the venture of building a larger and stronger machine that can be equipped with a power source heavy duty enough to propel a craft the size of the new one I have designed to the height and distance it will be capable of going."

"Certainly, you know that I have a keen interest in aeronautics, Mr. Dyer. But I don't have that natural ability you are blessed with to design and build flying machines," Mr. Redwine said. "I have friends who are interested in the subject but whether they will actually contract with me to enter a business enterprise of this sort is uncertain."

"Well, let your friends know about my strong desire to sell my project to someone. I'm positive this is going to be the movement of the future. I believe if a man, or a group of men, will get behind the building of it, they will have a valuable item to present to the world of commerce. It's not a small idea, I tell you. Building a machine that can reliably fly across many miles is a prize being sought all over the world right now. Sure, it's expensive to produce it, but the end result is worth far more than what it is going to cost.

"I know I really get wound up on this subject," Clark told him apologetically. "But I can see it so clearly and have no doubt about what the future holds for flying."

"When I hear you talking about it, Mr. Dyer, I know also that it is going to be a major means of transportation in the days to come. I will see what I can do to help get a buyer for you. Do you have a price in mind for it?"

"I feel like four thousand dollars would be a bargain price for it. I have put a lot of money into different phases of it down through the years, not to mention thousands of hours and risk of life in testing new designs of it."

"I might be able to find you a buyer at that price. I will let you know immediately if I do," Mr. Redwine said.

Clark rose and shook his hand. "I'm very grateful for your efforts on my behalf to get this project into good hands. It will do my heart good to know that someone is going forward to complete it. Let me know if there are questions raised by any interested person about it."

"I will certainly do so. You gentlemen keep warm in your travel back home. You're brave to be out on the road so far from home this time of the year," Mr. Redwine said.

They bade him farewell, boarded their wagon and turned toward home.

35. Medical Emergency

The weather was not so cold the first two days returning home from Gainesville, but the third day a drizzle of rain and sleet began to fall. A cloud cover caused the air to feel very chilly. Henderson didn't find it to be too uncomfortable, but Clark was plainly experiencing a great deal of discomfort, which concerned Henderson.

"Let's spend the night with Amy and John when we get to Cleveland, Papa," Henderson said.

"Yes, let's do. Maybe the sun will come back out tomorrow and it will be warmer."

But when they were a couple of miles from Amy and John's house, Clark suddenly slumped over in the wagon seat. Henderson brought the horse to a halt and pulled Clark into the wagon and wrapped him snuggly with a blanket over his coat and scarf. He helped him recline against a saddlebag.

"Papa, I will get to Amy's house as fast as I can. Hold on. We'll get you warmed up when we get there."

Clark made an incoherent sound, and Henderson was even more alarmed. He slapped the reins on the horse's neck and kept it at a trot the rest of the way to his aunt's house.

It was not yet five o'clock when they arrived, but the cloud cover made it seem much later. Henderson leapt from the wagon, ran to the door and knocked loudly. His Uncle John came to the door and didn't recognize him immediately.

"Uncle John, I'm Henderson. I have Papa in the wagon and he is sick. Can you come help me bring him in?"

"Yes, of course," he said, as he made out who it was. They both ran to the wagon. Clark was able to sit up only with assistance and they lifted him out, Henderson holding him under the arms and John holding him under the legs. They got him into the house and laid him on a davenport in the living room by the fire. Amy brought pillows and raised his head a little.

"Clark, it's me, Amy, your sister. You're going to be okay now. Will you drink some warm soup?"

Clark seemed to be trying to say, "No," but Amy rushed to the kitchen to get a cup of soup anyhow, believing that it would be good for him. When she brought it to him and placed it to his lips, he took a little in his mouth but didn't seem to be able to swallow it.

"John, go see if you can find Dr. Pendleton and bring him to see about Clark. Tell him that he looks like he might have had a stroke."

She removed his boots and found his feet to be ice cold. She warmed a flatiron beside the fireplace, wrapped it in a blanket and placed it at his feet. She got a night stocking and put it on his head, pulling it down over his ears. She got Henderson to help her shove the davenport closer to the fireplace. She took Clark's cold hands in hers and began to rub them gently as she whispered a prayer for his healing.

By the time John returned with the doctor in tow, Clark was sleeping, but the doctor tried to rouse him up enough to see if he was able to talk. He talked very loudly to him, asking him questions. Finally, Clark opened his eyes a little and looked at the doctor.

"Hello, Mr. Dyer. I'm Doc Pendleton and I want to know how you're feeling. Do you hurt anywhere?"

Clark shook his head in the negative.

"Can you say 'no'"?

Clark made only a garbled sound.

"Are you cold?"

Clark again weakly shook his head in the negative.

"Can you eat a little warm soup?"

Clark nodded affirmatively.

"Let's try putting some warm soup in a spoon and giving it to him," Dr. Pendleton told Amy. She quickly went to the kitchen and brought another cup of soup with a spoon.

The doctor asked Henderson to help him raise Clark up a little and he adjusted the pillows behind him. He dipped soup into the spoon and put it to Clark's mouth. Clark appeared to like the soup, but again he didn't seem to be able to swallow it, and Amy wiped his mouth with a cloth.

"Doctor, is there anything you can do for him?" she asked anxiously.

"Willow bark tea is about the only thing that might help him. If you have a kettle of water on the stove, bring it to a boil and I will measure out some bark to brew. The problem will be how to get the tea down him if he can't swallow. We will try to tilt his head back and spoon some down his throat. Maybe that will work."

Amy put extra wood under the stove eye beneath the kettle and lit the kerosene lamps as darkness descended. She waited impatiently for the tea to brew, all the while praying silently for the Lord's healing touch to be upon her brother.

When the tea was finally made, Henderson and John held Clark up while the doctor tilted his head backward and spooned the medicine down his throat. After each spoonful, they all waited nervously to see if it would stay down and breathed a sigh of relief each time it did. When Clark had kept several spoonsful down, they eased him back on the pillows and let him rest.

"Well, we can thank the Lord for being able to get that much medicine down him," the doctor said. "It encourages me that later we will also be able to succeed in slowly getting other liquids down him so he won't become dehydrated. You all keep an eye on him through the night, and I'll be back tomorrow to see how he is doing."

"We'd be happy to have you stay for supper with us, Doc," John said.

"No. Thank you for the offer, but I'll get on home. I surely hope it's not going to sleet any more tonight."

Amy went to the kitchen and put supper on the table. She was feeling a lot more hopeful that Clark would be better in the morning after he had had the medicine and would get a good night's sleep.

They sat down to eat and John asked the blessing, thanking the Lord also for providing help to Clark.

Amy was up several times throughout the night to check whether Clark was still sleeping and to put another stick of wood on the fire. She found that he had hardly moved each time she looked in on him. When she stepped in the next morning, he opened his eyes and gave her a faint smile.

"I hope you're feeling better, my brother," she said, coming around the end of the davenport to pat him on the arm.

He nodded a little.

"Could you eat some cream of wheat and drink some coffee?"

Again, he nodded.

Amy went to the kitchen and put the coffee on to boil. She started the cream of wheat cooking and a pan of tenderloin pork slices frying. She would make biscuits and fry eggs when the men had finished the feeding chores and were ready to eat. They would want gravy too, and that might be something Clark would be able to eat as well. She wanted so much for him to be able to swallow his food.

As soon as the coffee finished boiling, she poured a cup half full and added some cream. She spooned cream of wheat into a bowl and took both in to Clark. She placed them on a little table and pulled it over to the davenport.

"Do you think you can raise your shoulders up and scoot around if I help you? You'll be able to eat a lot better if you're sort of sitting up."

"Okay," he said, though his speech was quite slurred.

She pulled, he pushed and together they got him into a sitting position.

"See if you can pick up your spoon now," Amy said.

He reached shakily with his right hand and took hold of the spoon.

"That's great, Clark! You're stronger this morning!"

"Tired," he said. Then to Amy's surprise he added, "Hungry."

"Well, you just eat everything you want. I'll bring everything in the kitchen to you if you'll eat it. Let me hold your cream of wheat bowl up where you can get your spoon into it."

Despite being slow and a little shaky, he was able to dip the spoon into the cereal and lift it to his mouth without help. When he had eaten almost all of it, Amy asked if he wanted more. He shook his head. She handed him the coffee and he took it in his right hand. She noticed that he was not lifting his left arm.

"Can you raise your left hand?" she asked him.

She saw his fingers on that hand move, but he didn't lift his arm.

John and Henderson came in from the barn and were surprised to see Clark sitting up.

"Did you lift him up by yourself?" Henderson asked.

"He helped me by pushing as I was pulling. He can use his right arm and hand, but it looks like now his left side is unworkable. We can talk to the doctor about it when he comes today and see what he thinks about that."

"I'm very happy to see that he is able to sit up and eat." Lowering his voice, Henderson added, "Don't mention anything about us going home. He certainly doesn't need to travel right now."

Amy nodded in agreement.

"You and John come on and eat breakfast," she said. "I'm going to bring some gravy and biscuit to Clark and see if he will eat it."

When she brought it to him, he dipped a couple of pieces of biscuit into the gravy and ate it, but then handed the rest of it back to her.

"Do you want anything else? Some eggs? More cream of wheat? More coffee?" Amy asked him.

He shook his head negatively to all of her questions and slid down to rest his head on the pillows again. She helped him straighten himself out on the davenport and fluffed the pillows under his head.

"Take a nap until the doctor comes to check you out," she told him.

The doctor came shortly after noontime, and Clark was already awake when he came through the door.

"Well, look at you! Are you feeling better today, Mr. Dyer?"

"Yes," Clark answered.

"Amy told me that you're having trouble moving your left arm," the doctor said, taking hold of his left hand.

"Can you feel when I squeeze your hand?"

"Yes."

"Can you squeeze my hand?"

"No."

"You appear to have had a little stroke. Maybe you will be able to regain some use of your left arm as you exercise it. I think you should continue taking a little willow bark tea every day. It will help thin your blood and maybe prevent you from having further strokes."

The doctor turned to Henderson. "I would wait a couple of days for some better weather before traveling on over to Union County. Even then, keep your father as warm as possible. He doesn't need to have the stress of being chilled. Call me if he has any negative change in his condition while you're here."

The weather turned warmer two days later and Clark was urging Henderson to take him on home.

"Aunt Amy, heat some flatirons and wrap them in a wool blanket for Papa. Also, let's take a couple of your quilts and wool blankets to cover him with. He's going to worry himself sick if I don't take him home now."

They got him into the wagon and covered him warmly. Amy began to cry at seeing her brother so sickly and helpless.

"Clark, promise me that you will take care of yourself. I don't want you to get sicker again."

"I'll be careful to do what I can to keep well, honey. You do the same. God be with you till I see you again."

She gave him one last hug, and Henderson called "Giddap" to the horse and they were on their way.

Morena became concerned when Henderson and Clark hadn't returned from Gainesville within the usual timeframe for their trip. Because of Clark's bad heart and the cold winter weather, she was worried that something had gone wrong. They were at least two days past their usual return time. Matthew had stayed with her for more than a week since Clark left, and she knew his father would be wondering why he hadn't come home by now.

"Matthew, go home and tell your father that Clark and Henderson haven't come back from Gainesville," Morena said. "Tell him that I think somebody needs to go see about them."

Matthew hurried home and gave the message to Mancil, who agreed that somebody did need to check on them.

"You go back to your grandmother's house and stay with her until I bring word about Clark and Henderson."

He immediately saddled his horse and left to begin the search. He had traveled about three miles and was starting up the Logan Trail when he spotted Henderson's wagon coming down the mountain. He strained to see if Clark was sitting beside Henderson, but it looked like he was alone on the wagon seat. Feeling alarm at that, he urged his horse into a trot to reach him quickly.

"Where's Papa?" he asked as soon as he was within hearing distance.

"He is lying down in the wagon," Henderson said. "He got sick just as we were nearing Cleveland, and we spent a couple of days with Amy and John for him to regain some strength. Actually, it would have been far better if he hadn't started traveling again this soon, but you know him. He was getting very upset about coming home, so we bundled him up the best we could and I came on home with him. I'll tell you all about it when we get him home and beside a warm fire."

Mancil turned his horse around and started down the trail in front of Henderson's wagon.

They reached Clark's home in about an hour, and Morena ran out of the house upon hearing the horses. Like Mancil, her first look at the wagon with Henderson sitting on the wagon seat alone told her something was seriously wrong.

"Where is Clark?" she called to them.

"He's in the wagon, Mama. He's sick," Henderson said. "We'll bring him in. Fix a place for him to lie beside the fire and make him a cup of sassafras tea."

Morena ran back in the house and stoked the fire beneath the teapot. Then she hurried to the living room and got Matthew to help her shove the davenport near the fireplace. Henderson and Mancil came carrying Clark in and laid him on the davenport.

"Get a couple of pillows to put under his head," Henderson said. "Matthew, get the quilts and blankets in the wagon. Get the flatirons, too. Bring them here quickly."

They got Clark situated speedily and Morena pulled a chair up beside him.

"Clark, are you okay? Tell me what happened."

The strain of the cold weather and bumpy traveling in the wagon had weakened him again and he could barely speak.

"Will tell you later," he said in a raspy voice.

She reached for his hand and got the left one which was limp and cold. A cry escaped from her, "Oh, God, what has happened?"

Henderson wrapped his arms around her shoulders.

"Mama, he's had a stroke and can't use his left arm. I think he'll get better when we get him warm again. Be as calm as you can and just let him know that he's back home. You know him, he was struggling to get here as fast as he could."

Morena prayed to the Lord for strength and calmness to give comfort to him. She took both his hands and stroked them gently. "Honey, you're going to be all right. We'll have you warm and feeling better really quick." She turned to Mancil, "See if the sassafras tea is ready yet."

He went to the kitchen and found water boiling in the kettle on the stove. He poured the water over sassafras roots and covered the teapot to let it steep. His heart went out to his parents. They had been married such a long time and had been through so much together. If his father didn't pull through this, he didn't know how his mother was going to make it.

While the tea was steeping, he placed the flatirons next to the fire and laid extra quilts on chairs close to the fire to warm them for placing over his father.

As soon as the tea was ready, Mancil poured a cup half full and brought it to his mother. He and Matthew raised Clark up so he could drink, and Morena held the cup to his mouth for him. Clark also lifted his right hand to the cup and sipped the tea slowly. By the time he finished it, their hands had been warmed by the tea cup. Morena rubbed both his hands between hers and was encouraged as she felt the coldness dissipating.

"Sleepy," Clark said.

"Well, you just go right to sleep and have a good rest, Honey. You're home now. I'm going to wrap some flatirons in a quilt and put them at your feet. I love you and I'll be right here to see that you're warm and taken care of all night."

A smile crossed his lips and he relaxed into the pillows.

"Mama, I will get a mattress off one of the beds and bring it here for you to sleep on beside Papa," Henderson said. "Matthew can stay here another night with you until we see how Papa is doing tomorrow. Send him for me if you need anything. I love you, Mama. Don't worry. It will be okay. God will be with us."

Mancil and Henderson stepped outside and Mancil asked, "How bad off do you think Papa is?"

"I'm afraid he's pretty bad," Henderson said. "Maybe he would have had the stroke even if he hadn't gone to Gainesville. After all, he does have heart failure and has had it for several years. But his weakness and the extremely cold weather created a bad combination."

"I've decided I'm going to stay here tonight, too," Mancil said. "Will you go by my house and tell Rebecca that I'm staying and will see her tomorrow unless Papa is worse. I think we need to let the whole family know about his sickness. We probably shouldn't leave Mama here alone with him as long as he's this sick. We don't have to tell her that we're staying because we're worried. We can just treat it like a normal visit."

"I agree," Henderson said. "I'll be back to see about him tomorrow. I will surely be glad to get to sleep in my own bed tonight."

"I know you will."

36. Passing on the Dream

C lark awoke about four o'clock the next morning and couldn't figure out where he was. A fire was burning warmly in the fireplace. He looked around the room and spotted the mattress on the floor beside the davenport. In the semi-darkness it took him a few minutes to recognize that the woman asleep on the mattress was Morena.

Then he realized he was home. Joy flooded his heart. No more bouncing around in the back of a cold wagon. He wondered why he was on the davenport and how he got there. Then he realized he couldn't lift his left arm. As though remembering a bad dream, he started to recall being at Amy's house and a doctor saying he had had a stroke. Maybe it's not as serious as the doctor has judged it to be, he thought. Certainly, his head felt clearer this morning than it had in several days. He wondered if he could still walk.

So many thoughts racing through his mind brought the exhaustion back upon him, and he closed his eyes to slow the rushing torrent. In a few minutes, he was fast asleep again.

About six o'clock, Morena awoke and, like Clark, wondered for a few minutes where she was. But she quickly realized she was keeping watch over her sick husband and sat up to see how he seemed to be doing. In the early dawn light, he looked very peaceful in his sleep. The fire had burned low and she rose to put more wood on it. As she did so, suddenly a voice behind her spoke.

"Good morning, my love."

She spun around with her arms clasped to her breast, her eyes wide with fright. She recovered her breath when she recognized it was Clark speaking. She ran to his side.

"Honey! Are you better this morning?"

"Yes, I am. You're a sight for sore eyes."

"But you could hardly speak last night. What happened?"

"I don't know. I don't even remember how I got here. I was so cold and tired."

Mancil was awakened by the voices and jumped out of bed and ran into the living room to see what was going on.

"Is everything all right in here?" he asked.

"Your papa nearly scared me to death while I was putting wood on the fire," Morena said.

"Did he have an attack of some kind?"

"No," Morena laughed. "He said, 'Good morning, my love' and I didn't know he could speak that strongly because his voice was so weak last night. I didn't know where it had come from."

Mancil looked at his father and had to agree with Morena. He was very much improved from the fellow he and Henderson had carried into the house from the wagon last night.

"How do you feel, Papa?"

"I feel much better. The past few days have been like a bad dream."

"Well, I'm certainly happy to see you doing so much better this morning."

"I am wondering if I'm still able to walk since the stroke has numbed my left arm. Maybe you can help me stand up and see if I can."

"Well, here I am in my long johns," Mancil said, holding his hands out in mock exasperation. "What do you say we wait until I get my clothes on before I start taking you for a walk?"

"Oh, sure. That will be fine," Clark said with a slight smile.

Mancil got into his clothes quickly and came back in the room.

"I will get Matthew to come help me in case I need him," he told his mother.

"Maybe we should wait until I cook breakfast and eat before you try. He's pretty weak since he hasn't had much food for the past few days," Morena said. "Why don't you raise him up and see how much movement he has in his left leg while he's sitting."

"That's a good idea, Mama," Mancil said.

He stepped over to his father. "Papa, let me help you sit up and we'll see if you can move your leg."

Mancil helped him raise his head and shoulders then placed pillows behind him.

"Mama, take his feet and legs and turn them as I push his shoulders around to rest against the back of the davenport."

When they had him in a sitting position, he rubbed his left leg and tried to lift it. He couldn't pick it up but he could move his foot slightly.

"Well, it looks like I'm hobbled right now," he said sadly. "With an arm and leg that barely budge, there's not a lot that I can do."

"We'll help you work on it, Papa," Mancil told him. "I have the wheelchair you built for Mr. Collins to use when he broke both his legs during the War Between the States. The family gave it back to me after he passed away. I'll bring it for you to use now. You will probably be able to get around using a walking cane before long. Don't be discouraged."

The family worked with Clark daily to help him regain strength in his arm and leg. As Mancil had predicted, he began to have some limited use of his limbs again, but it didn't appear he would recover enough function in them to be mobile any time soon without using the wheelchair. He was starting to become discouraged about his disability, and especially that he could no longer work on his flying machine. The cold, dreary January weather kept him sitting by the fireplace day after day, inwardly fretting over his condition. He tried to hide his depression from the family because he knew they were doing all they could to keep his spirits up.

There was one thing he should do, Clark decided. He should write out his wishes regarding what was to be done with the flying machine when he passed on.

"Morena, bring me a pencil and paper," he said to her as she sat near him beside the fireplace, spinning wool into yarn.

She stopped spinning and studied him a moment. His ill health and unhappiness were causing deep lines to form in his face. She knew in her heart what he was thinking and rose to go and get the requested items. If this would relieve any of his anxiety, that would be good. Returning, she handed him the paper and pencil.

"Let me roll your wheelchair into the kitchen so you can put your paper on the table to write."

She rolled him to the dining table and laid the paper in front of him. She lifted his left arm and laid it on the edge of the paper to hold it in place while he wrote.

"Do you want me to help you?" she asked.

"I'll try to do it myself, but if I can't I will let you do it for me."

He began to write: I offered to sell my flying machine, the patent and my plans for the larger machine to John Redwine in Gainesville for four thousand dollars. We did not sign any agreement for this but it is my wish that if I should die before these items are sold to anyone and Mr. Redwine should want to purchase them afterward that our agreement be honored by my family. He signed and dated the paper.

He felt a wave of exhaustion and called to Morena, "I'm finished. Come and roll me back to the fireplace."

She looked at the short instructions he had written. "Were you able to write everything you wanted to?"

"Yes, that's it. I just wanted to make sure it was understood what I wanted to be done about the agreement I made for selling the flying machine."

She rolled him back to his favorite place beside the fireplace. She leaned down and placed her head against his.

"Honey, don't give up hope that you're going to get better. I don't know what I would do without you."

"I don't really feel good, Morena. I've tried very hard to keep my spirits up, but I think I may not get over this."

Tears began to roll down Morena's cheeks. She wiped them quickly, not wanting him to see her crying. She tried to speak with a steady voice.

"You'll get better when spring comes and you can go outside again," she said reassuringly.

"Maybe so."

He lifted his right hand and enfolded her hand that was resting on his shoulder. He hated to think of her having to live without him. They had been joined heart to heart for so many years, he knew it would be extremely difficult for her to walk alone.

"Honey, just always know that my spirit is with you whether I'm physically present or not. And just as we look to the day when we'll be reunited with our parents in heaven, you can look forward to the day when you and I will also be together again. Whichever one of us goes first, it will only be a temporary absence."

As he spoke, Morena's tears began flowing profusely and she sank to the floor with her head on his knees. She knew these were his farewell words to her and her heart was breaking.

When Morena went to awaken him the next morning, she found that he had quietly passed into eternity during the night. She sat on the side of the bed and whispered to him through her tears, "Honey, walk beside me as you promised. If you don't, I won't be able to make it. Oh, how I will miss you! I'm glad your sadness is over and you can walk in the sunshine again, but I wish I could be with you, too. I will always love you."

Matthew had spent the night with them, and she went to awaken him and send him to tell the family that his grandfather had passed away.

"Tell your father that he and your uncles can make all of the funeral arrangements. I am not feeling strong enough to handle it myself. One of you tell Rena to come and get me. I want to stay at her house for a while. Matthew, you have been such a blessing to us through these weeks, staying here and doing chores for us through your grandfather's sickness. God will bless you for your goodness to us."

"Grandma, I have been happy being here and helping. I will be sure that Aunt Rena gets the message to come and get you. We'll take care of everything here at your house while you're gone. I know you're sad, Grandma. I love you."

He gave her a warm hug and hurried up the trail to deliver the sad report.

Clark had been gone two months and signs of spring were beginning to appear once again in Choestoe. Farmers were beginning to harrow the fields in preparation for planting their crops. Morena had come back to her home to live and the grandchildren were alternating weeks to come and stay with her.

Morena had told the family about Clark's verbal agreement with John Redwine for selling the flying machine to him, and they agreed that it was the best course of action for them to take.

She had written Mr. Redwine to inform him of Clark's death and his desire for her to honor his verbal agreement.

One day she saw a covered wagon coming up the trail toward her house with two men seated at the front. She supposed that it might be someone coming to discuss sale of the machine. When it pulled into the yard, she went onto the porch.

"Hello," one of the men said. "I'm John Redwine. This is my brother, Phillip."

"Hello. I'm Morena Dyer, Clark Dyer's widow."

"I am so sorry for your loss, Mrs. Dyer. Your husband was a brilliant and kind man. I know he will be missed."

"Thank you. Yes, we miss him terribly."

"I came to talk to you about Mr. Dyer's flying machine. May we take a look at it?"

"Yes. It's in the workshop up there. Go ahead and take a look."

The two men alighted from the wagon and walked up to the shop. They spent about half an hour inside, then came back to the house and knocked on the door.

When Morena opened the door, John Redwine said, "Mrs. Dyer, that's a fine machine. We do want to buy it along with the patent and his plans for the larger flying machine. Is it your understanding that he wanted four thousand dollars for everything?"

"Yes. That is what he said."

"That's the amount we brought with us and I will step out here to the wagon and get it and a bill of sale for you to sign. Do you have the patent and the plans for the larger machine?"

"Yes. I will get them for you."

Morena went to the bedroom and opened the bottom drawer of the chest where Clark had kept the patent the entire seventeen years since he had received it. She recalled how very proud he had been the day he opened the letter from the U.S. Patent Office and pulled out the official patent issued in his name. The plans for his new machine lay neatly folded beside it. He had wanted to apply for a patent for it too, but he never was able to accumulate enough money to file the application.

Morena shook off the sad reminiscence of Clark's hindered plans. He had always been able to maintain hope, and she must do likewise.

She brought the documents out to Mr. Redwine, and he was waiting on the porch with the money and the bill of sale. She invited the men to come in and count the money and sit at the table where she would sign the paper. With the sale officially completed, Mr. Redwine shook her hand.

"Thank you, Mrs. Dyer. It was a pleasure meeting you and transacting the sale. I'll drive the wagon up to the shop and load the flying machine in it, then we'll be on our way."

She watched from the porch as they got the machine loaded and started down the trail with it. The place felt empty with the last traces of Clark's forty years of work on the flying machine completely gone.

"Lord, let someone build the machine Clark envisioned and let him rejoice in heaven when it sails through the skies as he knew it would one day."

Descendants of Clark Dyer—Howard Dyer, great-grandson, and Kenneth Akins, great-great-great-grandson—looking toward Rattlesnake Mountain. (*The Times*, Gainesville, GA, Sunday, March 16, 1980; used by permission)

Epilogue

After his death in 1891, Clark's patent and airplane were sold, and the family had only stories handed down orally about his invention for more than a century. There were no cameras or newspapers available in the mountain area during Clark's lifetime to support the stories told about his flights, and searches conducted by historians Kenneth Akins and Robert Davis in 1980 failed to produce any documentation about his airplane. It isn't known what journals or notes of the dates and details of his flights may have been kept by Clark or some of his family members; but if there were any, they apparently were turned over to the buyers of the airplane in 1891.

In 1994, Sylvia Dyer Turnage, great-great-granddaughter of Clark Dyer, published a book titled *The Legend of Clark Dyer's Remarkable Flying Machine* in an effort to prevent the story from becoming lost to future generations. The book contained a ballad she wrote about the legend as she had heard it told by family members all of her life, and it is included at the end of this Epilogue.

Then, in 2004, Steven and Joey Dyer, great-great-great-grandsons of Clark Dyer, located the patent for the airplane in the U.S. Patent & Trademark Office. The patent office had loaded old patent files onto their site on the Internet, which included Patent No. 154,654 issued on Sept. 1, 1874, to Micajah Dyer of Blairsville, Ga. for an "Apparatus for Navigating the Air." Later, about two dozen 1875 newspapers published in Georgia towns outside of the mountain area and in other states across the country were discovered that included articles about Clark Dyer's invention.

In recent years, the family has made efforts to gain recognition for the inventor. They make the case that this is not merely for family heritage, but because it is an important historical event for the state, indeed for the nation, since no one preceded Clark Dyer in inventing a flying machine that was capable of controlled flight.

In 2006, Georgia Highway 180 East was named the Micajah Clark Dyer Parkway resulting from the introduction of a Resolution by State Representative Charles Jenkins and signed by Governor Sonny Perdue to honor Clark for his invention.

In 2009, the Micajah Clark Dyer Foundation, Inc. was formed to educate the public about Clark's aeronautical accomplishments. In the same year, Sylvia Dyer Turnage published another book, this one titled *Georgia's Pioneer Aviator, Micajah Clark Dyer* to provide the new information discovered after her 1994 book was published. By this time, the story of Clark's invention had been presented before many schools and civic groups; and many magazines, newspapers, radio and television shows had told about the historical event as well.

The original stone that marked Clark's grave was replaced in 2010 with a granite monument. The original soapstone markers that were located at his and his wife's graves were embedded in the monument, as well as graphical drawings of the airplane from his patent. His grave is located in the Old Choestoe Baptist Church Cemetery, just off the Micajah Clark Dyer Parkway.

Since Clark's aircraft is no longer available for viewing, the Foundation felt the need for constructing a model from the drawings and descriptions in his patent. In 2012, Jack Allen, a retired machinist from Delta Airlines, agreed to build a scale model of the airship, and it is now on display in the Union County Historical Society Museum in the Old Courthouse on the Square in Blairsville, Georgia. The original smaller model of the machine that was crafted by Brian Paquette in 2006, and which had been on display in the Historical Society's museum for several years, was moved to the Union County Public Library to be included in an exhibit there.

The Foundation has other goals they hope to accomplish during the coming years to give recognition to Clark for his invention, including naming the Blairsville Airport the Micajah

Clark Dyer Airport; getting a full-size replica of the plane built and displayed in a museum for him at the airport; inducting him into the Georgia Aviation Hall of Fame at Robins Air Force Base in Warner Robins, Georgia; continuing to make presentations about this portion of history to schools and civic groups; and attracting an entertainment organization to produce a movie of his life.

The ballad below was written ten years before Clark's patent was discovered by family members.

The Ballad of
Clark Dyer's Flying Machine

1.

Clark Dyer stood on a mountainside
Watching the birds in flight.
Said to himself, why couldn't I
Make a machine so light
It could soar o'er the mountains high
Taking a man to places far
Giving him speed, lifting his soul,
Making him feel like a star.

Chorus:
Clark wanted to fly, fly, fly
He wanted to fly.
He wanted to fly way up high
Into the sky.
He wanted to fly, fly, fly
He wanted to fly.
He wanted to fly until the day
That he died.

2.

Some of his neighbors thought he was mad
To believe that a man could fly.
"Can't he see that the weight is too much to lift
A man and his craft to the sky?"

But Clark was wise in designing ways
To do many things others could not do.
He felt in his heart that a craft could be made
To fly like he wanted to.

3.

So he set to work with his primitive tools
Making the frame with pine
Covered it o'er with canvas he made
From the shucks of his corn so fine.
He built some rails on the mountainside
For his plane to increase its speed
It rose from the earth and took to the air
A sight to behold indeed.

4.

Now the years were advancing on Clark Dyer
Yet he kept working on that rig,
Trying to find some way to get power
That would move a craft so big.
But the Grim Reaper came and took him away
With his work unfinished yet.
We honor him still for his pioneering feat,
And his memory we'll not forget.

-- Sylvia Dyer Turnage, ©1994

Appendix

Additional information can be found at:
www.micajahclarkdyer.org

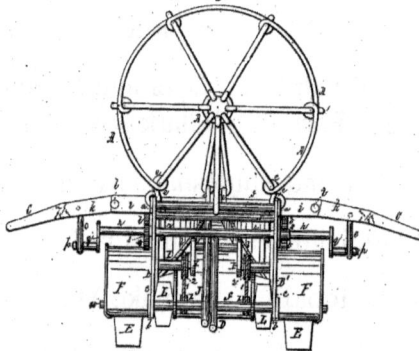

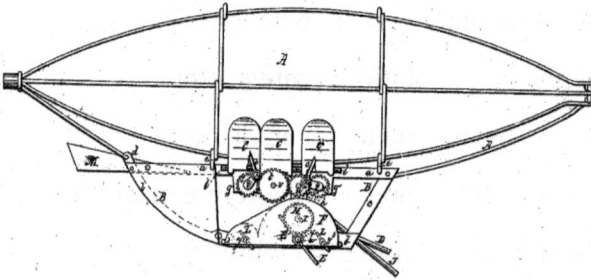

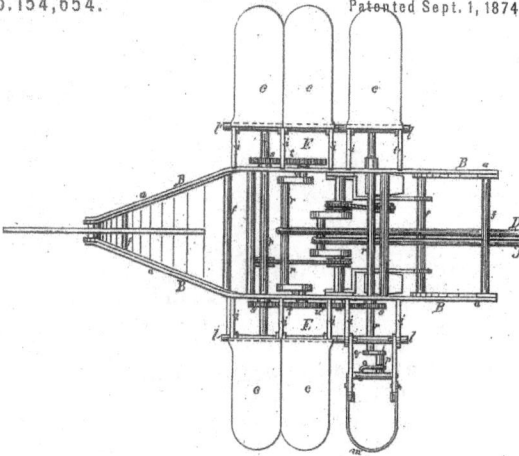

M. DYER.

Apparatus for Navigating the Air.

No.154,654.

Fig.3.

Patented Sept. 1, 1874.

Fig.4.

Witnesses.

Inventor

Sheet 2 of drawings-U.S. Patent 154,654, dated Sept. 1, 1874

The Latest Flying Machine.

[From the Gainesville (Ga.) Eagle.]

We had a call on Thursday from Mr. Micajah Dyer, of Union county, who has recently obtained a patent for an apparatus for navigating the air. The machine is certainly a most ingenious one, containing principles entirely new to aeronauts, and which the patentee confidently believes have solved the knotty problem of air navigation. The body of the machine in shape resembles that of the fowl, an eagle, for instance, and is intended to be propelled by different kinds of devices, to wit: Wings and paddle-wheels, both to be simultaneously operated, through the instrumentality of mechanism connected with the driving power. In operating the machinery the wings receive an upward and downward motion, in the manner of the wings of a bird, the outer ends yielding as they are raised, but opening out and then remaining rigid while being depressed. The wings, if desired, may be set at an angle so as to propel forward as well as to raise the machine in the air. The paddle-wheels are intended to be used for propelling the machine, in the same way that a vessel is propelled in water. An instrument answering to a rudder is attached for guiding the machine. A balloon is to be used for elevating the flying ship, after which it is to be guided and controlled at the pleasure of its occupants.

How It Feels to be Hanged.

A Paris newspaper gives this extract from the notes of a young fellow who tried to com

Athens Banner-Watchman
Athens, Georgia, Tuesday, April 28, 1885

AERIAL NAVIGATION.

A Union County Man Invents a Flying Machine.

BLAIRSVILLE, April 19, 1885.—
Mr. Editor: Mr. Clark Dyer, of this
county, thinks that he has succeeded in making an air ship. He says
that he can sail through the air, but
has not means to get material of
sufficient strength to complete the
machine. He has had it patented,
and he now only lacks capital to
complete the machine, and he wants
capitalists to engage in the matter.
He believes if he can get scientific
men to examine his machine, they
will come forward and aid in its
completion, and as you have always
been a friend to science and inven-
tors, he wants you to take such steps
as in your judgment will bring his
invention before the public generally.

Mr. Dyer is in earnest, and has,
no doubt, made something that he
can propel through the air. If you
think it possible to get men of
means to investigate the matter, I
would be pleased if you would do
so.

Mr. Dyer has worked thirty years
on his machine. He is not crazed,
but is in dead earnest, and confidently believes that he has solved
the problem of aerial navigation.
He is not a crank nor
a fanatic, but is a good, quiet citizen and a successful farmer.
Anything you may think proper to
say or do in his behalf, I will take
as a personal favor. He does not
want to humbug anybody. After
examination, if he has nothing he
wants nothing. JOHN M. RICH.

The Athens Banner-Watchman, weekly edition, No. XLIV, Vol.
XXXI, dated Tuesday, April 28, 1885

Micajah Clark Dyer and Morena Owenby Dyer

Portrait of Clark Dyer by Artist Doris Durbin

Clark's home place looking from Rattlesnake Mountain

Rattlesnake Mountain where Clark launched his aircraft

Field at the foot of Rattlesnake Mountain where Clark landed
his flying machine

Stink Creek at the foot of Rattlesnake Mountain

Map of Union County, Georgia

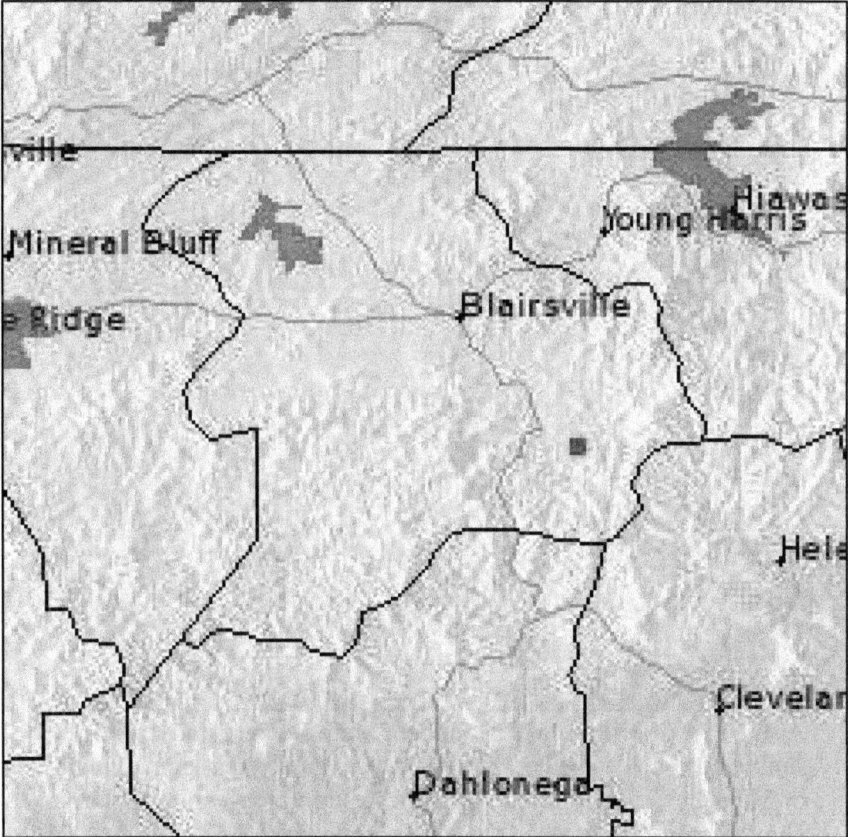

The red dot on the Map of Union County, Georgia (Blairsville is the county seat), marks the site of Clark Dyer's flight. A portion of Tennessee and North Carolina state lines share Georgia's northern boundary.

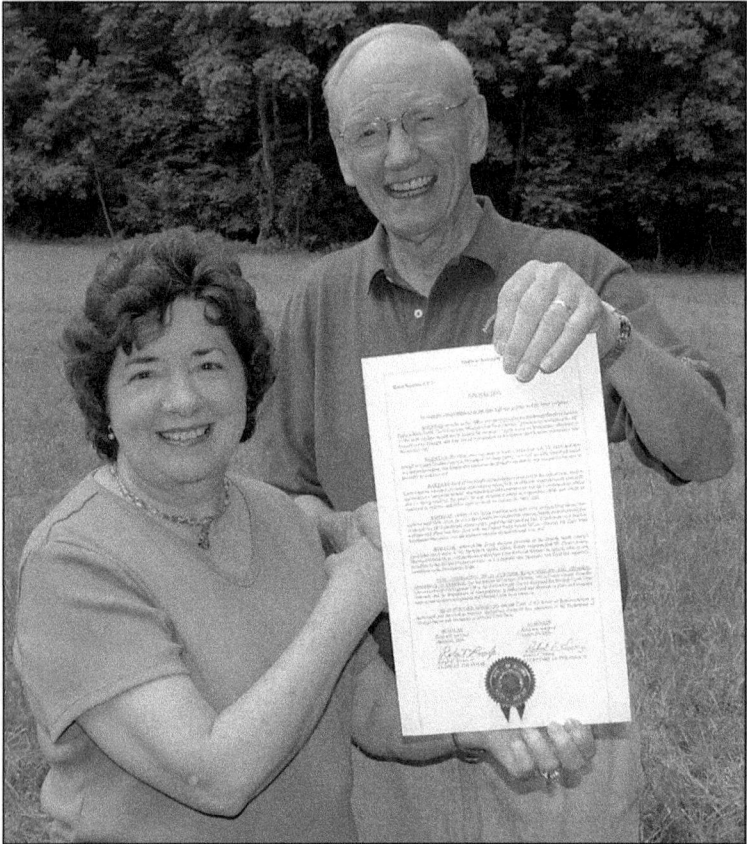

At the foot of Rattlesnake Mountain, State Representative Charles Jenkins presents to Sylvia Dyer Turnage, great-great-granddaughter of Micajah Clark Dyer, the signed Resolution for naming Georgia Hwy. 180 East "Micajah Clark Dyer Parkway."

MICAJAH CLARK
DYER PARKWAY
PIONEER AVIATOR
· 1822 – 1891 ·

Governor Sonny Perdue with Dyer family at the State Capitol in 2006 for the ceremonial signing of Resolution to name Georgia Hwy. 180 East the Micajah Clark Dyer Parkway.

Governor Perdue hoists James Micajah Cooper, great-great-great-great-grandson of Micajah Clark Dyer, into the air.

A monument to Clark Dyer was erected at his gravesite in the Old Choestoe Baptist Church Cemetery just off the Micajah Clark Dyer Parkway (GA Hwy. 180-E). Embedded in the monument are the old hand-carved soapstone headstones that were originally at Clark's and Morena's graves and negatives of two of Clark's drawings that were included in his patent for an "Apparatus for Navigating the Air" issued Sept. 1, 1874.

The back side of the monument above.

Jack Allen, a retired Delta Airlines machinist, built a scale model of Clark Dyer's apparatus for navigating the air. He crafted every piece of the model, including all of the working parts—gears, pulleys, paddle wheels—with superb skill.

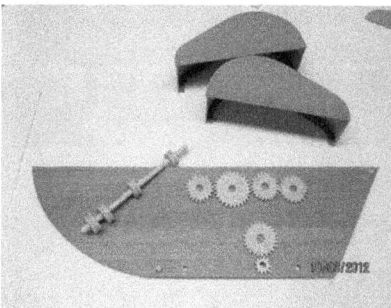

The completed model built by Jack Allen is on display at the Union County Historical Society's Museum in the Old Courthouse on the Square in Blairsville, Georgia. The mirrored bottom of the display case permits a view of the interior working parts. The wings are left uncovered to show how they operate.

Union County Commissioner Lamar Paris hangs a display of a 39-cent U.S. Postage stamp on the wall in the Union County Courthouse. The stamp was designed to honor pioneer aviator Micajah Clark Dyer and features one of the drawings taken from his U.S. Patent granted on September 1, 1874 for an Apparatus for Navigating the Air.

Commissioner Paris signs a Proclamation declaring Sept. 1, 2006 "Micajah Clark Dyer Day in Union County."

Dyer Family Crest

𝕯𝖞𝖊𝖗

This is an Occupational name from the Old English word D E A G R E
meaning "dyer"--- one who either processes the dye or one who dyed the
cloth. Chaucer's Canterbury Tales says:
 " An haberdasher, and a carpenter,
 A webbe, a deyer, and a tapiser."
 The "webbe" spun cloth as a spider spins webbs and a "tapiser"
was one who made a kind of figured material for wall decorations.
 Henry le (the) Deghar lived in Somersetshire in 1260 and Robert
le Deyare lived in Worcestshire in 1275. Alexander Dygher lived in
Sussex in 1296 and Henry le Dyer lived in Derbyshire in 1327. Richard le
Dyr of Norfolk, was a rector of Fincham church in 1333. Ricardus Dir was
on the 1379 Yorkshire poll tax rolls.
 John Dyer was given a clearance to sail for New England in 1634 on
the ship Chrystean. Anamieh and John Dyer sailed for Virginia in 1635.
John Dyes and John Dyers came in 1642 and 1653 respectively. Robert and
John Dye came in 1650 and 1655 respectively

TERRERE NOLO TIMERE NESCIO

Created July 6, 1678
Arms: On a Chief indented, gu
Crest: out of a ducal, Cornet or,

 a goat head, arg, armed, gold.
Motto; Tierrere - Nolo- Timere- Nescio

 " I do not fear,
 I will not affright."

Dyer Crest
Certificate of Authentication

Children of

Micajah Clark Dyer
(b. July 13, 1822 in Pendleton District, SC
d. January 26, 1891 in Union County, GA)
and
Morena Elizabeth Owenby Dyer
(b. December 24, 1819 in Rutherford, NC
d. September 27, 1892 in Union County, GA)

Clark and Morena were married on July 23, 1842.

They are buried in the Old Choestoe Baptist Church Cemetery in Union County, GA

Clark was the son of Sarah Elizabeth ("Sallie") Dyer and John Meyers who never married. He was raised by his maternal grandparents, Bluford Elisha Dyer, Jr. (born about 1775, died in 1847) and Elizabeth Clark Dyer (died in 1861), whom he called Pa and Ma. He called his birth mother by her nickname, Sallie.

Morena was the daughter of Robert H. and Matilda ("Mattie") Owenby.

As was typical of the mountaineers, Clark and Morena had a large family of 9 children—7 boys and 2 girls. They were:

1. Jasper Washington Dyer ("Jasper")
Born March 19, 1843, died Aug. 15, 1913, buried in Pine Top Baptist Church cemetery. Married Emaline ("Emma") Lance who was born July 30, 1843, died Jan. 9, 1915.

2. John M. Dyer (who became a minister)
Born Dec. 12, 1845, died around 1900, buried in the Dyer family cemetery in Choestoe. Married Elizabeth Ann Sullivan.

3. Andrew Henderson Dyer

Born Jan. 16, 1848, died Oct. 16, 1903, buried in Old Liberty Baptist Church cemetery. Married Adeline Sullivan who was born Nov. 18, 1848, died Aug. 6, 1921.

4. Marcus Lafayette ("Fate") Dyer

Born June 7, 1850, died July 7, 1921. Married Clarissa Wimpey who was born June 16, 1850, died Dec. 30, 1921, buried in Towns County.

5. Cynthia ("Cindy") C. Dyer

Born June 15, 1852, died Oct. 25, 1917, buried in Lumpkin County. Married John P. Smith who was born June 7, 1846.

6. Mancil Pruitt Dyer

Born May 16, 1854, died March 17, 1916; buried in Pine Top Baptist Church cemetery. Married (1) Rebecca Jarrard and (2) Margaret M. Twiggs, born Aug. 2, 1871, died May 28, 1949.

7. Robert F. Dyer

Born April 19, 1856, died around 1900, buried in a family cemetery on the old Rouse Waldrop place (formerly owned by the Dyers) in Arkaquah District, Union County. Married Elizabeth Fortenberry.

8. Morena Elizabeth ("Rena") Dyer

Born Sept. 14, 1859, died April 9, 1903, buried in Old Choestoe Baptist Church cemetery. Married James A. Wimpey who was born Aug. 15, 1856, died Feb. 15, 1894.

9. Johnson Benjamin Dyer

Born July 7, 1861, died about 1885, buried in Pine Top Baptist Church cemetery. Married Mary Hunter.

Children of

Bluford Elisha Dyer, Jr. and Elizabeth Clark Dyer
(Clark's Grandparents)

1. Sarah Elizabeth ("Sallie") Dyer
Born about 1805, married Eli Townsend about 1825. Birth mother of Micajah Clark Dyer II.

2. Hezekiah Dyer
Born about 1807, married Sarah Dalton on December 25, 1832.

3. Girl, name unknown
Born about 1809. (Called "Amy" in novel; husband called "John.") She was married and did not move to Choestoe with her parents in 1833.

4. Girl, name unknown
Born about 1811. (Called "Mary" in novel, husband called "Pete.") She was married and did not move to Choestoe with her parents in 1833.

5. Lucinda ("Lucy") Dyer
Born about 1813, married James Crow, first Sheriff of Union County, Georgia, in 1832.

6. Bluford Elisha Dyer III
Born about 1814, married Mary Younce about 1832, later moved to Tennessee.

7. Micajah Clark I ("Cager") Dyer
Born about 1817, married Harriett Logan on June 25, 1839.

8. Elijah ("Lige") Dyer
Born about 1819, married Mary ("Polly") Kettle.

■ *Micajah Clark Dyer II (grandson)*
Born July 13, 1822, married Morena Owenby on July 23, 1842.

9. James Marion ("Jimmy") Dyer
Born about 1823, married Eliza Ingram on June 18, 1846.

10. Melinda Dyer
Born about 1827, married William B. Harkins on Dec. 5, 1844.

11. Matilda Dyer
Born about 1830, married Francis M. "Frank" Swain on Aug. 6, 1855.

12. Bluford Lumpkin ("Lump") Dyer
Born about 1832, married Ruth Turner on Feb. 8, 1854. He was Sheriff of Union County in 1866.

Children of

Robert H. Owenby
and
Matilda ("Mattie") Owenby
(Morena's Parents)

1. Jonathan Owenby, born Dec. 24, 1816.

2. Morena Owenby, born Dec. 24, 1819.

3. Barbara Owenby, born Feb. 4, 1822.

4. Ann Owenby, born Feb. 21, 1824.

5. Alfred Owenby, born Sept. 25, 1826.

6. Joseph Owenby, born Dec. 4, 1828 (twin).

7. John Owenby, born Dec. 4, 1828 (twin).

8. Ephraim Owenby, born Dec. 1, 1831.

9. Powell Owenby, born July 24, 1833.

10. Matilda Salina Owenby, born May 18, 1837

11. Eleanor Eveline Owenby, born Feb. 1840.

12. Sarah Owenby, born 1842.

13. Robert Anderson Owenby, born Nov. 1845.

www.ingramcontent.com/pod-product-compliance
Lightning Source LLC
Chambersburg PA
CBHW052031090426
42739CB00010B/1857